Cambridge Elements ≡

Elements in the Philosophy of Mathematics
edited by
Penelope Rush
University of Tasmania
Stewart Shapiro
The Ohio State University

MATHEMATICS AND METAPHILOSOPHY

Justin Clarke-Doane
Columbia University

CAMBRIDGE
UNIVERSITY PRESS

CAMBRIDGE
UNIVERSITY PRESS

University Printing House, Cambridge CB2 8BS, United Kingdom

One Liberty Plaza, 20th Floor, New York, NY 10006, USA

477 Williamstown Road, Port Melbourne, VIC 3207, Australia

314–321, 3rd Floor, Plot 3, Splendor Forum, Jasola District Centre, New Delhi – 110025, India

103 Penang Road, #05–06/07, Visioncrest Commercial, Singapore 238467

Cambridge University Press is part of the University of Cambridge.

It furthers the University's mission by disseminating knowledge in the pursuit of education, learning, and research at the highest international levels of excellence.

www.cambridge.org
Information on this title: www.cambridge.org/9781108995405
DOI: 10.1017/9781108993937

First published 2022

A catalogue record for this publication is available from the British Library.

ISBN 978-1-108-99540-5 Paperback
ISSN 2399-2883 (online)
ISSN 2514-3808 (print)

Mathematics and Metaphilosophy

Elements in the Philosophy of Mathematics

DOI: 10.1017/9781108993937
First published online: June 2022

Justin Clarke-Doane
Columbia University
Author for correspondence: Justin Clarke-Doane, jc4345@columbia.edu

Abstract: This Element discusses the problem of mathematical knowledge and its broader philosophical ramifications. It argues that the challenge to explain the (defeasible) justification of our mathematical beliefs ("the justificatory challenge") arises insofar as disagreement over axioms bottoms out in disagreement over intuitions. And it argues that the challenge to explain their reliability ("the reliability challenge") arises to the extent that we could have easily had different beliefs. The Element shows that mathematical facts are not, in general, empirically accessible, contra Quine, and that they cannot be dispensed with, contra Field. However, it argues that they might be so plentiful that our knowledge of them is unmysterious. The Element concludes with a complementary "pluralism" about modality, logic, and normative theory, highlighting its revisionary implications. Metaphysically, pluralism engenders a kind of perspectivalism and indeterminacy. Methodologically, it vindicates Carnap's pragmatism, transposed to the key of realism.

Keywords: philosophy of mathematics, knowledge, realism, objectivity, pluralism

ISBNs: 9781108995405 (PB), 9781108993937 (OC)
ISSNs: 2399-2883 (online), 2514-3808 (print)

Contents

Introduction   1

1 Self-evidence, Analyticity, and Intuition   3

2 Observation and Indispensability   13

3 Connection, Contingency, and Pluralism   22

4 Modality, Logic, and Normativity   35

Conclusions   45

References   46

Introduction

The *Twin Primes Conjecture* says that there are infinitely many prime numbers, *p*, such that *p + 2* is also prime.[*] As for this writing, it is an open question whether this conjecture is true or false, although most experts believe that it is true. If the question is settled, it will be settled by proof.

What is a proof? It is basically an argument that convinces experts of the claim proved. More fully, it is an argument that convinces experts that there exists a *formal* proof of that claim.[1] A formal proof of *S* is a finite sequence of sentences, each of which is either an axiom or follows from the previous sentences by a rule of formal inference, the last line of which is *S* itself.

Whether a proof is sound is widely supposed to be a mind- and language-independent matter. It is supposed to be independent of us whether the logic used is correct and whether the argument is valid in that logic. This is *logical realism. Mathematical realism* is the view that it is independent of us whether the (nonlogical) axioms are true (and that some are non-vacuously true). Their combination says that we do not make up the logical or mathematical facts.

Mathematical realism raises a question: How do we know that the axioms are true?[2] Even if knowledge of the logical axioms is intelligible (a matter to which we return in Section 4), mathematical axioms are not just (first-order) logical truths. Consider the *Axiom of Choice* (*AC*). This says that if t is a disjointed set not containing the empty set, $\emptyset$, then there exists a subset of $\cup$t whose intersection with each member of t is a singleton. In symbols: $(t)[(x)[x \in t \rightarrow (\exists z)(z \in x) \& (y)(y \in t \& y \neq x \rightarrow \sim (\exists z)(z \in x \& z \in y))] \rightarrow (\exists u)(x)(x \in t \rightarrow (\exists w)(v)[v = w \longleftrightarrow (v \in u \& v \in x)])]$. It is consistent with standard mathematics, minus *AC*, that *AC* is false if standard mathematics is consistent. Universes in which *AC* fails are studied and deeply understood – unlike, say, universes in which squares are circles. So there is nothing "unintelligible" about choiceless mathematics, in any ordinary sense. How, then, do mathematicians know that *AC* is true?

---

[*] Chapters 1 and 3 draw on Chapters 2 and 5, respectively, of my book, *Morality and Mathematics* (Oxford University Press, 2020). For overviews of much of the technical material discussed here, see the following additional Cambridge Elements: *Set Theory* by John Burgess, *Gödel's Theorems* by Juliette Kennedy, and *Foundations of Quantum Mechanics* by Emily Adlam.

[1] Although this is the standard view, it can be questioned. See De Toffoli [2021]. (Of course, no one should be under the illusion that mathematics *as practiced* simply consists of deducing theorems from axioms. See Harris [2015] for a lovely portrayal of the experience of pure mathematical research.)

[2] Logical realism bears on the more general question of how proofs supply mathematical knowledge. For instance, how do we reliably determine that there exists a formal proof on the basis of the (informal) arguments that convince experts (setting aside the question of how we know that the axioms of feasible length are true)? This is not obvious because formal proofs of mainstream theorems are typically too lengthy to be humanly comprehensible. See Gaifman [2012, 506].

The difficulty can sound generic. There is also the question of how scientists know their laws. How do physicists know the *Dirac Equation* or geologists know the theory of plate tectonics? There is nothing unintelligible about the failure of these claims either. The difference is that in these cases, we have the beginning of an answer. Electrons and the Earth's crust leave marks on the world to which our nervous systems respond. We bear no relevant physical relations to the likes of sets. So no story like this suggests itself in connection with our knowledge of *AC*.

To be clear, it is not that mathematical cognition is beyond the reach of science. It is an established subject in cognitive psychology.[3] But we must distinguish the study of mathematical belief acquisition from the study of the *correlation* between our mathematical beliefs and the facts. Science is illuminating our beliefs about sets (and, especially, natural numbers). But it has been conspicuously silent on how they relate to the mathematical facts.

It is tempting to dismiss the problem as stemming from an unwarranted "platonism" about mathematical entities. If platonism is just mathematical realism – that is, the view that the *Twin Primes Conjecture* is either true or false, independent of what anybody says or believes – then the problem does stem from this. But platonism in this sense is difficult to discharge. Our best theories of the physical world are up to their ears in mathematics (Section 2). So absent a way to "factor out" those theories' mathematical commitments, the view that the mathematical truths depend on us would seem to imply that the physical truths do too. For example, the (time-dependent) Schrödinger Equation of quantum mechanics tells us how the state vector of a physical system changes with time. How could this express an independent fact if there are not any independent facts about vectors? Or consider the banal claim that some of our scientific theories are at least *consistent*, that is, do not (classically) imply a contradiction, independent of what anyone says or believes. This claim turns out to be a simple arithmetic claim whose negation (for typical theories) is consistent if elementary arithmetic itself is consistent.[4]

So the question of how humans acquire knowledge of independent mathematical facts is pressing. This Element clarifies the problem, sketches a solution, and discusses its import for philosophy more generally, including modal metaphysics, (meta)logic, and normative theory.

---

[3] See Butterworth [1999], Carey [2009], De Cruz [2006], Dehaene [1997], Pantsar [2014], and Relaford-Doyle and Núñez [2018] for work on the psychology of number concepts. See Marshall [2017] and Opfer *et al.* [2021] for a critical discussion of the relevance of work like this to the problem of mathematical knowledge.

[4] Technically, the claim is a Π_1 claim, that is, a claim of the form "for all natural numbers, Φ," where Φ is a formula with only bounded quantifiers.

1 Self-evidence, Analyticity, and Intuition

Knowledge is justified and non-coincidentally true belief (where specifying the relevant sense of "coincidence" is the so-called Gettier Problem). So the problem of explaining our knowledge of the axioms, partitions into two. First, there is the problem of explaining the (defeasible) justification of our belief in the axioms, what I call the *justificatory challenge*. Second, there is the problem of explaining our belief's non-coincidental truth, that is, the *reliability challenge*. Let us begin with the first.

1.1 Two Kinds of Axiom

What are the axioms of mathematics? There are two varieties. On the one hand, there are axioms that just speak of their class of models. These are *structural* axioms. For example, a mathematical *group* is any set that is closed under a binary operation satisfying the axioms of associativity, identity, and invertibility. We may also stipulate that the operation is commutative. In that case, the group is said to be *Abelian*. But there is no nonverbal question as to whether the *Axiom of Commutativity* itself is true. It is true of Abelian groups and false of the others.

The situation is prima facie different with *foundational* axioms, like those of set theory, type theory, category theory, and arithmetic. Foundational axioms are, roughly, those on the basis of which one can carry out metatheoretic reasoning. For instance, already in (first-order) *Peano Arithmetic* (*PA*), one can formulate claims about the consistency of theories and prove *relative consistency* results. One can prove, say, that if *Zermelo–Fraenkel* (*ZF*) set theory is consistent, then so is *ZFC* + Cantor's *Continuum Hypothesis* (*CH*); where *ZFC* is *ZF* plus *AC*, and *CH* says that there is a bijection between every uncountable subset of the real numbers and all of them, or, equivalently, given *AC*, that the cardinality of the real numbers is the next greatest after that of the natural numbers. This is written: $PA \vdash \mathrm{Con}(ZF) \rightarrow \mathrm{Con}(ZFC + CH)$. Arithmetic axioms do not seem to be structural – just about their class of models – because there seems to be a nonverbal question as to whether metatheoretic claims like the aforementioned are true. Indeed, Gödel's *Second Incompleteness Theorem* says that if *PA* is consistent, then it cannot prove that it is, written $PA \nvdash \mathrm{Con}(PA)$. Nor, thankfully, do we have that $PA \vdash {\sim}\mathrm{Con}(PA)$. So $PA + \mathrm{Con}(PA)$ and $PA + {\sim}\mathrm{Con}(PA)$ are both consistent if *PA* is – just like group theory with the Axiom of Commutativity and group theory with the negation of that axiom. But the question of whether arithmetic is consistent cannot be dismissed like the question of whether the Axiom of Commutativity for groups is true! There either is or is not a natural number that codes a proof (in classical logic) of "$0 = 1$", from the axioms of *PA*. Or so it seems. If this were not the case – if, for instance, $PA + \sim \text{Con}(PA)$ were really analogous to Abelian group theory – then there would also be no nonverbal question as to what counts as finite (since a model of $PA + \sim \text{Con}(PA)$ disagrees with us about this), what a formula is, and even what a theory, like *PA*, consists in.

Such considerations only take us so far. They do not show that there is a nonverbal question of whether characteristic axioms of set theory, like *AC*, are true, for example. Indeed, a view to which we return in Section 3.5 says, roughly, that nonstructural foundational axioms are limited to those of (first-order) arithmetic [Weaver 2014, Ch. 30]. But many realists deny that we should draw the line at arithmetic (Koellner [2014], Van Atten and Kennedy [2009], Woodin [2010]). Is it really just a verbal question whether, for any disjointed set not containing $\emptyset$, there is a subset of $\cup$ x whose intersection with each member of x is a singleton? What about the claim that there is a so-called *Inaccessible Cardinal (IC)*? This implies new arithmetic results. $ZFC \nvdash \text{Con}(ZFC)$ (if *Con(ZFC)*), but $ZFC + IC \vdash \text{Con}(ZFC)$. So, arguably, belief in *IC* is presupposed by belief in *ZFC*.[5]

Whatever we include among the structural and foundational axioms, there are axioms that cleanly qualify as neither. Tarski's axioms for first-order geometry do not have the flavor of the Axiom of Commutativity for groups (Tarski [1959]). Prima facie, they have an intended subject, Euclidean space, of which they could be wrong. But those axioms are also not foundational, in that one cannot carry out metatheoretic reasoning in the theory.[6] The *Parallel Postulate*, which says, informally, that two straight lines intersecting another so as to make less than a 180° angle on one side intersect on that side, will serve as a key example in Sections 3 and 4. A debate over it would be misconceived, like a debate over the Axiom of Commutativity for groups. But, unlike group theory, this is not because geometry is about its class of models. It is because, if geometric reality exists, it is rich enough to afford an *intended model* of the postulate and its negation.[7]

---

[5] "Arguably" because the assumption of *IC* is stronger than the assumption that there is a model of *ZFC* (which is equivalent, by *Soundness* and *Completeness* to Con(*ZFC*)). That is, $ZF + IC$ is stronger than $ZF + \text{Con}(ZFC)$.

[6] The theory is decidable and complete and so, by Gödel's theorems, cannot even interpret *Robinson Arithmetic* (i.e., *PA* minus all instances of the *Induction Schema*).

[7] The distinction between structural and foundational axioms is similar to Shapiro's distinction between algebraic and non-algebraic ones, although he appears to think that the distinction is exhaustive. See Shapiro [1997, 41 and 50]. Likewise, Balaguer [2001] distinguishes between mathematical domains in which our intentions are exhausted by the (first-order) axioms that we adopt from those in which they are not. This is different from the distinction above if our intentions about a domain can transcend any recursive axiomatization while failing to interpret arithmetic.

1.2 Self-evidence

Despite being a relatively fringe area of pure mathematics, set theory is of special philosophical interest. While it has only one nonlogical predicate, $\in$, the claims of all other branches of mathematics can be *interpreted* in it. Those claims can be understood as claims about sets in disguise. It does not follow that all mathematical entities are *really* sets (Benacerraf [1965]). It follows that if the axioms of set theory are consistent, then so are our other mathematical theories.[8]

Could the axioms of set theory, and of all other areas of mathematics, including arithmetic, be consistent but false (or vacuous)? Not if consistency is understood standardly, as a claim about proofs, or set-theoretic models. One could take the notion of consistency as primitive (an idea to which we return in Section 2.3). But on what basis might we believe that, say, set theory is thus consistent? Perhaps the standard answer is: on the basis that it is true, and truth implies consistency (Frege [1980/1884, 106]; Woodin [2004, 31])! But this answer, in tandem with the assumption that mathematical claims are true independent of us, implies mathematical realism.

What, then, explains the justification of our belief that the axioms are true? In other words, why is it *rational* or *reasonable* for us to believe those axioms? A common answer outside of the philosophy of mathematics is that "[a]xioms are mathematical statements that are *self-evidently* true" [Greene 2013, 184, italics in original]. This is perhaps defensible in rudimentary cases.[9] Consider the *Axiom of Extensionality*, which says that if "two" sets have the same members, then they are really one and the same (the converse is a logical truth in first-order logic with identity). In symbols: $(x)(y)(z)[(z \in x \longleftrightarrow z \in y) \rightarrow (x = y)]$. Set theory without Extensionality has been explored (Friedman [1973]; Hamkins [2014]; Scott [1961]). But this axiom is often taken to be some kind of truism about sets. Similarly, the *Axiom of Pairing* says that for any "two" (perhaps not distinct) sets, there is another containing just those two. That is: $(x)(y)(\exists z)(w)$ $[w \in z \longleftrightarrow (w = x \lor w = y)]$. This is also difficult to deny – though it is unclear that any existential statement, even conditional on the existence of other objects, could be self-evident.

---

[8] Whether set theory, rather than another theory, or no theory, can serve as a "foundation" for mathematics in any of the myriad senses that term have been discussed will be irrelevant. It is certainly not unique in interpreting mathematics (see, e.g., Tsementzis & Haverson [2018]). However, it is canonical in this respect, so I focus on it for concreteness.

[9] Authors rarely say exactly what they mean by "self-evident." But the idea seems to be that *P* is self-evident when, if one understands *P*, one is thereby (defeasibly) justified in believing *P* (the "thereby" would require explication).

However, Extensionality and Pairing do not imply the existence of a single set![10]
Set theory gets going with the *Axiom of Infinity*, written: $(\exists y)(((\exists x)(x \in y \,\&$
$(z)(z \notin x) \,\& \,(x)(x \in y \to (\exists z)(z \in y \,\& \,(w)(w \in z \longleftrightarrow (w \in x \lor w = x)))))$.[11]
This says that there is an inductive set – that is, a set that (according to the usual
definitions) includes *0* and includes the number $n + 1$, whenever it includes n. It is
hard to see the point of calling the claim that something *infinite* exists "self-
evident" (Mayberry [2000, 10]).[12] Other axioms are still more doubtful. Consider
the *Axiom of Replacement*. This is a schema, not a single axiom. It says that for any
set, z, and *any formula*, Φ, such that, for every $t \in z$, there is exactly one x with
$\Phi(t, x)$, there exists a set that contains just those things, x, for which $\Phi(t,x)$
holds for some $t \in z$. Formally: $(a)[(u)(v)(w)(u \in a \,\& \,\Phi(u, v) \,\& \,\Phi(u, w) \to$
$u = w) \to (\exists y)(x)(x \in y \longleftrightarrow (\exists t)(t \in a \,\& \,\Phi(t, x)))]$, where u, v, w, and y
are not free in $\Phi(t, x)$. This has important consequences for set theory, like the
Reflection Principle (to which it is actually equivalent in the context of the other
axioms), which says that if a formula is true of the set-theoretic universe, V, then
it is already true in an initial segment, V_α, of it. The Axiom of Replacement is
even needed to prove that the number $\omega + \omega$ exists. But it also implies the
existence of outrageously huge sets (though they are tiny for set theory!). Of a
relatively small such set, κ, Boolos, laments: "Let me try to be as accurate,
explicit, and forthright about my belief about the existence of κ as I can … I …
think it probably doesn't exist" [1999, 121].

Finally, consider, again, *AC*. In the context of the other axioms, *AC* is
equivalent to the claim that every set is well-orderable (totally orderable so
that every non-empty subset of it contains a least element). Thus, *AC* ensures
that the set of real numbers, *R*, has a well-order. But what is that order? It cannot
be the standard order, since there is no least real number in any open subset of
real numbers, like (0, 1). In fact, it is consistent with *ZFC* (if that is consistent!)
that there is no *definable* well-order on *R* at all – that is, no well-order specified
by a formula, no matter how lengthy and baroque. Even if *AC* is *true*, it is not
self-evident!

Needless to say, if typical axioms like Infinity, Replacement, and Choice, are
not self-evident, then neither are speculative extensions of them, contra the

---

[10] It is a classical logical truth that there is an x such that $x = x$, since domains are *defined* to be non-
empty. But Extensionality and Pairing give us nothing beyond this, an assumption which can,
anyway, be dropped by adopting a free logic.

[11] *PA* and *ZF* minus Infinity plus its negation are actually bi-interpretable (if the *Axiom of
Foundation*, to be discussed, is stated as a scheme of ∈-induction). So Infinity is essential to
set theory, as opposed to arithmetic. See https://math.stackexchange.com/questions/315399/
how-does-zfc-infinitythere-is-no-infinite-set-compare-with-pa

[12] The claim that there is an inductive (infinite) set must be clearly distinguished from the claim that
there are infinitely many things. Set theory, minus Infinity, proves the latter, but not the former.
The former proves Con(*PA*).

rhetoric of some set theorists. Consider Gödel's *Axiom of Constructibility,* $V = L$. Let $P_{Def}(A)$ refer to the set of all subsets of A definable in the structure $\langle A, \in \rangle$ by first-order formulas with parameters in A. Then $V = L$ says that every set lies in the following hierarchy obtained by transfinite recursion on the ordinals: $L_0 = \emptyset, L_{\alpha+1} = P_{Def}(L_\alpha)$ and $L_\gamma = \cup_{\alpha < \gamma} L_\alpha$ for limit γ (Gödel [1990/1938]). Is $V = L$ true? The dominant narrative, originating with Gödel himself (see his 1947 work), is that $V = L$ must be false because it settles undecidables – especially "large" large cardinal axioms – in the wrong way (Maddy [1997, Pt II, § 4]; Magidor [2012]; Woodin [2010, 1]). But Fontanella points out that Gödel's "feeling that [V=L's] consequences would be implausible is not unanimously shared" [2019, 32]. Indeed, Jensen writes, "I personally find [$V = L$] a very attractive axiom" [1995, 398]. He continues, "I do not understand ... why a belief in the objective existence of sets obligates one to seek ever stronger existence postulates [large cardinal axioms]. Why isn't Platonism compatible with the mild form of Ockham's razor ...?" [1995, 401].[13] Devlin thinks that $V = L$ "is ... a natural axiom, closely bound up with what we mean by 'set' .... [and] tends to decide problems in the 'correct' direction" [1977, 4]. And Eskew queries, "The axiom $V = L$ ... settles 'nearly all' mathematical questions .... [I]t can be motivated by constructivist views that are still widely held today .... [A] wealth of powerful combinatorial principles ... follow from ... $V = L$ .... [So] why hasn't there been ... a stronger push to adopt it as a[n] ... axiom for mathematics?" [2019).][14]

1.3 Analyticity

So appeal to self-evidence does not afford a satisfying answer to the justificatory challenge. How else might we explain the justification of belief in the axioms? Another prominent proposal is that the axioms are *analytic*, "a system of tautologies, the basic elements of which are true by virtue of the meanings of

---

[13] See Arrigoni [2011] for an explication and defense of Jensen's position.

[14] Consequences of $V = L$ that are said to be particulary counterintuitive (besides that there does not exist a so-called *Measurable Cardinal*) include there is a definable but nonmeasurable set of reals, and the *Diamond Principle* holds. Gödel [1947] contains further arguments against the axiom. On the other hand, Fraenkel, Bar-Hillel, and Levy [1973, 108–109] contain additional arguments supporting $V = L$, and Simpson compares skepticism about large cardinals (the larger of which imply that $V = L$ is false) to (rational) religious skepticism in his [2009]. Friedman quips, "[some s]et theorists say that $V = L$ has implausible consequences ... [and] claim to have a direct intuition which allows them to view these as so implausible that this provides 'evidence' against $V = L$. However, mathematicians [like me] disclaim such direct intuition about complicated sets of reals. Many ... have no direct intuition about all multivariate functions from N into N" [2000]! Arrigoni and Friedman emphasize that criteria of success and intuitiveness vary, and that "ZFC $+ V = L$ ... is fruitful in consequences, furnishes powerful methods for solving problems and introduces the concept of *constructability*, important throughout set theory" (Arrigoni & Friedman [2012, 1361, italics in original]).

the terms used" [Singer 1994, 8]. In light of Quine [1951a], most philosophers are careful to distinguish epistemic from metaphysical versions of this proposal.[15] The *metaphysical* version says that the meaning of the term "$\in$" somehow *makes it the case* that the axioms are true. This is hard to even understand. How could a meaning make a fact? The *epistemic* version says that it is "part of the concept of $\in$" that standard axioms hold, and those of us with that concept are, therefore, defeasibly justified in believing those axioms (at least assuming that we are justified in believing that there are any sets at all).

Supposing for the moment that the notion of epistemic analyticity is in good order, it is doubtful that standard axioms are so analytic. First, it is hard to imagine a compelling argument that it is just "part of the concept of $\in$" that standard axioms hold, given that some theorist actually denies them. Consider the *Axiom of Foundation* (or *Regularity*). This schema says that for any formula, Φ, if there is a set that satisfies Φ, then there is a *minimal* x that does – an x such that Φ and no $y \in x$ such that Φ. In symbols: $(\exists x)\Phi \rightarrow (\exists x)[\Phi \,\&\, (y)$ $(y \in x \rightarrow \sim\Phi^*)]$ (where Φ does not contain y and Φ^* is just Φ but contains y wherever Φ contains free occurrences of x). This is equivalent to a *Principle of Set-theoretic Induction*, according to which, if Φ is a formula such that, whenever all members of x satisfy Φ, x does too, then every set satisfies Φ. *Foundation* is widely alleged to be the foremost example of a nontrivial analytic axiom (Boolos [1971, 498], Shoenfield [1977, 327]). It is just part of what we mean by "$\in$" that every set is formed at some stage of a transfinite generation process via the powerset and union operations, beginning with $\emptyset$ – so that no set contains itself, and there are no infinitely descending chains of membership, for example. This "platitude" is equivalent to Foundation, given the other axioms. It says, if $V \neq L$, that all sets lie in a liberalized version of Gödel's L, the *Cumulative Hierarchy*: $V_0 = \emptyset$, $V_{\alpha+1} = P(V_\alpha)$, and $V_\gamma = \cup_{\beta<\gamma}V_\beta$, for limit γ, where $P(x)$ is the *ordinary* powerset operation (i.e., the set of *all* subsets of the set x, even those that are not definable in the structure $\langle x, \in \rangle$). But far from being beyond dispute, many doubt the *coherence* of the resulting "iterative conception of set" (*ICS*)! What, after all, could "formation" and "generation" mean when these terms are applied to the likes of (pure) sets (Ferrier [Forthcoming], Potter [2004, § 3.3])? Rieger complains: "[*ICS*] does not embody a philosophically coherent notion of set. There is a coherent constructivist position .... There is also a coherent anti-constructivist

---

[15] See Boghossian [2003] for the distinction.

position .... But [*ICS*] is an uneasy compromise between these two: it pays lip-service to constructivism without really meaning it" ([2011, 17–18].[16]

Even if it were just "part of the concept of $\in$" that standard axioms hold, however, epistemic analyticity is a suspect idea. If we were worried that some sets fail to occur at any V_α, then under the assumption that it is "part of the concept of $\in$" that all sets do, we should just worry *that our concept of set is not satisfied*. Maybe instead of sets, there are only set-like things, which are similar to sets except that some fail to live in any V_α (because they are, say, self-membered). Epistemic analyticity makes justification too cheap. For any claim of interest, S, consistent with the other claims that we believe, we could be justified in believing S simply by enriching our concepts! Of course, if every consistent concept of set – or, more carefully, theory in the language of first-order set theory – were satisfied (in a class model, under a face-value Tarskian satisfaction relation), then we might be able to rule out the worry that ours is not. But, if that were the case, then the whole project of seeking out the "true" set-theoretic axioms would be misconceived. *Every consistent set-theoretic sentence that was not a logical truth would be like the Parallel Postulate* (understood as a claim of pure mathematics). By Gödel's Second Incompleteness Theorem, this includes (a coding of) the claim that *PA* is consistent, if it is.

1.4 Reflective Equilibrium

So the axioms of set theory seem to be neither self-evident nor analytic in a useful sense. Is there any other way to explain the justification of our belief in them? Russell proposes what is perhaps the canonical way. He writes, "We tend to believe the premises because we can see that their consequences are true, instead of believing the consequences because we know the premises .... But the inferring of premises from consequences is the essence of induction; thus the method in investigating the principles of mathematics is really an inductive method, and is substantially the same as the method of discovering general laws in any other science" [1973/1907, 273–4]. Russell's proposal is that, first, the *epistemic* priority of mathematical principles is opposed to their *logical* priority. Although we deduce theorems from axioms, we are justified in believing the axioms because we are justified in believing the theorems that they imply, rather

---

[16] See also Azcel [1988, Introduction]. Advocates of the so-called logical conception of set, such as Quine [1937] and [1969], reject the Axiom of Foundation too. Quine's *New Foundations* (*NF*) for mathematical logic proves the existence of a universal set, which contains itself. (The relative consistency of *NF* is still officially an open problem. However, experts appear to be converging on the view that it is consistent even relative to quite weak theories. See: https://mathoverflow .net/questions/132103/the-status-of-the-consistency-of-nf-relative-to-zf)

than the other way around. Second, the theorems that justify us in believing the axioms need not be self-evident or analytic. They need only be initially plausible, or *intuitive*.

Russell's method prefigures *reflective equilibrium*, championed by Goodman and Rawls. Rawls writes, "Although … various judgments are viewed as firm enough to be taken provisionally as fixed points, there are no judgments of any level of generality that are … immune to revision" [1974, 8].[17] An attraction of the method is that it analogizes the epistemology of mathematics to that of empirical science, which is better understood. Gödel stresses that "the axioms need not … be evident in themselves, but rather their justification lies (exactly as in physics) in the fact that they make it possible for these 'sense perceptions' to be deduced" [1990/1944, 121]. But neither Russell nor Gödel distinguishes the justificatory and reliability challenges. The analogy at most holds for the former. Benacerraf complains, "there is a *superficial* analogy …. [W]e 'verify' axioms by deducing consequences from them concerning areas in which we seem to have more direct 'perception' (clearer intuitions). But we are never told how we know even these, clearer, propositions" [1973, 674, italics in original]. Field clarifies that "we [can] grant … that there may be positive reasons for believing in [select theorems]. These … might involve … initial plausibility …. But Benacerraf's challenge … is to … *explain how our beliefs about these remote entities can so well reflect the facts about them*" [1989, 26, my emphasis].

We discuss the reliability challenge in detail in Section 3. For the present, even the idea that the *justification* of our mathematical beliefs can be explained in analogy with the justification of our empirical scientific ones is tendentious. The problem is that there is *disagreement over the data to be accounted for* in the mathematical case that has no apparent analog in the empirical one.[18]

Consider a paradigmatic disagreement over an empirical scientific theory, the theory of dark matter. Those who reject the hypothesis of dark matter, like Milgrom [2002, 45], and propose amendments to Newtonian gravity do so in order to account for *the same data*.[19] They do not disagree over *it*. But disagreement in the foundations of mathematics seems *characteristically* to

---

[17] See also Goodman [1955, 63–64].

[18] This is why comments like the following are too quick. "Many realists … take the epistemological challenge to be one about … epistemic justification … .And they reply in the obvious ways … by showing that their favorite theory of epistemic justification in general nicely applies to the case of [mathematical] beliefs … .[T]his is not a promising way of understanding … the epistemological challenge … . [W]hatever your theory of epistemic justification, it is hard to see any special difficulties applying it to [mathematical] beliefs [Enoch 2009, 2]." (Enoch is actually talking here about normative beliefs, although the more general context is both normative and mathematical ones.)

[19] For details, see Milgrom's online overview here: http://ned.ipac.caltech.edu/level5/Sept01/Milgrom2/paper.pdf See Merritt [2020] for a philosophical discussion of Milgrom's program.

bottom out in conflicting intuitions – or what Jensen calls "deeply rooted differences in mathematical taste" [1995, 401]. We have already discussed clashing intuitions in set theory. For example, while most set theorists hold that $V = L$ resolves questions in the wrong way, others hold that it "tends to decide problems in the 'correct' direction" (and still others claim to have no intuition one way or the other about its consequences). Expert intuitions diverge in "core" areas, like analysis and number theory, too. Weyl maintains: "in any wording [the Least Upper Bound Axiom of the calculus] is false" (quoted in [Kilmister 1980, 157]), and Feferman [1992] follows him in this (sketching an analog to analysis without it).[20] Nelson begins his 1986 work, "The reason for mistrusting the induction principle [of *PA*] is that it involves an impredicative concept of number .... A number is conceived to be an object satisfying every inductive formula" [1986, 1]. So that concept is circular. Surely, though, at least *Robinson Arithmetic* (*PA* minus *all instances* of *Induction*) is sacrosanct? On the contrary, Zeilberger writes, "I am a platonist ... [but] I deny even the ... axiom that every integer has a successor .... " [2004, 32–3].[21] (Note that Weyl and Nelson, at least, accept classical logic. So their objections do not stem from disagreements over logic.)[22]

Of course, some empirical scientific disagreements bottom out in conflicting observations too. Observation is theory-laden, and perceptions can be impaired. Two pathologists looking at the same biopsy sample may disagree about the extent of dysplasia it harbors. The point is that such cases are the exception. Unlike disagreements in the foundations of mathematics, disagreements over empirical scientific theories do not seem to be *primarily attributable* to disagreements over the data.[23]

It is tempting to respond with poll numbers. Koellner tells us that "[*Projective Determinacy* (*PD*), which is inconsistent with $V = L$] has gained wide

---

[20] This is not a coincidence. Set-theoretic and "core" principles are often objectionable for the same reasons. For instance, Weyl rejects the *Least Upper Bound Axiom* because it is *impredicative*. It defines an object in terms of a totality to which it belongs, which Weyl thinks is incoherent (in mathematical analysis, at least). The *Subsets Schema*, $(z)(\exists y)(x)(x \in y \longleftrightarrow (x \in z \& \Phi))$, to be described in Section 2.1, does the same thing. Whenever the condition Φ refers to the powerset of z in this schema, it defines a subset z_Φ in terms of a set, $P(z)$, to which z_Φ belongs.

[21] Nelson sometimes demurs from asserting that every natural number has a successor as well, at least in connection with what he calls "actual" (or "genetic") numbers. See his [1986, 176] and [2013].

[22] Zeilberger does not say whether he accepts classical logic. For more on disagreement over axioms, see Forster [Forthcoming], Fraenkel, Bar-Hillel, and Levy [1973], Maddy [1988a & 1988b], and Shapiro [2009].

[23] The following caveat from another context is called for. "Diagnosing a clash of intuitions ... will typically involve attempting a careful hermeneutic reconstruction of the underlying dialectic, designed to reveal that the dispute rests ultimately with certain ... premises that one side finds intuitive and the other does not. Any such reconstruction is bound to be controversial .... [W]hereas many philosophers agree that some questions boil down to ... differences in intuition, there is considerable disagreement as to exactly which questions those are [Mogensen 2016, 24]."

acceptance by the set theorists ... who know the details of the constructions and theorems involved in the case that has been made for *PD*" [2013, 21–22]. But, even if true, it is hard to see what this could show. First, empirical scientists' judgments are tested against a world that bites back. There are no comparably robust means by which to calibrate conflicting mathematical intuitions. If anything, intuitions seem to track sociological factors, such as who one studied with and where one went to graduate school [Cohen 1971, 10].[24] Martin remarks that "For individual mathematicians, acceptance of an axiom is probably often the result of nothing more than knowing that it is a standard axiom" [1998, 218]. This is just the kind of correlation on which the epistemology literature has focused (Mogensen [2016]). It is like the observation that had we gone to a different graduate school, we would have believed, say, epistemological externalism, instead of internalism.

Second, the *relevant* group to poll would presumably consist of those who actually work on the disputed axioms and related problems – not just those who know the details of the constructions.[25] But, as every philosopher knows, specialist knowledge tends to turn "something so simple as not to seem worth stating" into "something so paradoxical that no one will believe it" [Russell 1918, 514]. Fraenkel, Bar-Hillel, and Levy identify "far-going and surprising divergence of opinions and conceptions of the most fundamental mathematical notions, such as set and number" among those working on foundational questions [1973, 14]. And Bell and Hellman [2006] begins, "Contrary to the popular (mis)conception of mathematics as a cut-and-dried body of universally agreed upon truths ... as soon as one examines the foundations of mathematics one encounters divergences of viewpoint ... that can easily remind one of religious, schismatic controversy" [64]. Unfortunately, heretic mathematicians are not like flat-earthers.[26]

A key question for Russell's answer to the justificatory challenge is, thus, whether intuitions' variability precludes them from (defeasibly) justifying, in the way that observations are supposed to justify. *Phenomenal conservativists* may deny that it does (Bengson [2015]; Chudnoff [2013]; Huemer [2005]; Pryor [2000]). We have to start somewhere. Where else but with plausibility judgments? However, those with even slightly reliabilist sympathies will claim that plausibility cannot be enough. Plausibility judgments must at least be tenuously

---

[24] Indeed, Koellner studied with the most vocal proponents of *PD*, like Hugh Woodin.

[25] Devlin draws on an analogous distinction when he writes: "Currently I tend to favour $[V = L]$ .... At the moment I think I am in the majority of informed mathematicians, but the minority of set theorists ..." [1981, 205].

[26] Forster jokes, "for people who want to think of foundational issues as resolved ... [standard axioms provide] an excuse for them not to think about [them] any longer ... . To misquote Chesterton "If people stop believing in [standard] set theory, they won't believe nothing, they'll believe anything" [Forthcoming, 15].

reliable symptoms of the truths. Consider a believer in *AC* and a believer in ~*AC*. Forster says, "The current situation with *AC* is that the contestants have agreed to differ" [Forthcoming, 72]. Evidently, "the contestants" have robust intuitions and beliefs in equilibrium.[27] Could their intuitions still (defeasibly) *justify* their belief in either *AC* or ~*AC*, despite their opposite deliverances? The answer apparently turns on the extent to which justification can come apart from reliability.[28]

Even if conflicting intuitions could *defeasibly* justify, *knowledge* of their variability might *undermine* whatever justification they afforded. If this were so, then the justificatory challenge would be moot. The upshot would be the same as if intuitions did not justify. We should be agnostics as to whether *AC* in the absence of a reason to suspect that those with opposite intuitions are mistaken, *which is independent of the question of whether AC*. We return to the question of how knowledge of others' intuitions might bear on our own when discussing the reliability challenge in Section 4.

1.5 Conclusion

I have discussed the justificatory challenge and responses to it in terms of self-evidence, analyticity, and reflective equilibrium. The first two responses do not stand up to scrutiny. The third response, due to Russell, has the virtue of analogizing the justification of our mathematical beliefs to that of our empirical scientific beliefs, which is more tractable. However, the analogy falters to the extent that disagreements over axioms turn on disagreements over the data to be accounted for, while disagreements over empirical laws tend not to.

Perhaps, though, Russell's analogy to empirical science does not go far enough. Maybe our mathematical beliefs are not only justified by data that is systematized in the way that observations in science are. Maybe, appearances notwithstanding, they are literally justified by observations. Maybe *pure* mathematics is an empirical science. I turn to this proposal now.

2 Observation and Indispensability

I have argued that the canonical response to the justificatory challenge (the challenge to explain the defeasible justification of our mathematical beliefs)

---

[27] It is sometimes suggested that all disagreements over *AC* bottom out in disagreement over classical logic (with detractors rejecting classical logic in favor of intuitionistic logic). But, as Forster [Forthcoming, Ch. 7] illustrates, this is not so.

[28] "Apparently" because this appearance will be complicated in Section 3.5.

falters insofar as mathematical intuitions vary in a way that observations do not. However, there is a revisionary account of the justification of our mathematical beliefs, according to which they are justified by observations, much like ordinary claims of theoretical empirical science. Let us consider this account.

2.1 The Web of Belief

The view that mathematics is an empirical science suggests itself in elementary cases. Mill [2009/1882] argues that our belief that *1 + 1 = 2* is inductively confirmed by observations of physical pairs. The problem is to extend an empirical story beyond grade school arithmetic. Quine [1951a, section VI] does this. Instead of arguing that mathematical facts are themselves observable, he argues that they are *scientific postulates*, like hypotheses about electrons and gluons. He writes, "Objects at the atomic level and beyond are posited to make the laws of macroscopic objects, and ultimately the laws of experience, simpler .... [T]he abstract entities which are the substance of mathematics ... are another posit in the same spirit ... neither better nor worse except for differences in the degree to which they expedite our dealings with sense experiences" [1951a, 42].[29] This is supposed to explain how our belief in core claims of standard mathematics, like the Axiom of Infinity, could be justified. Like our belief in gluon confinement, our belief that there is an inductive set is implied by the best systemization of our observations. Sets are like gluons.

Of course, sets are not *metaphysically* like gluons. Pure sets lack mass-energy, quantum number, and other physical attributes. Quine's claim is that sets are *epistemologically* like gluons. As Marcus puts it, "It is one of Quine's great achievements to notice that the [epistemological] access problem in the philosophy of mathematics becomes obsolete once we recognize that ontological commitment is a matter of formulating theories rather than grounding each individual claim in sense experience or [Gödelian] rational insight" [2017, 51].

In fact, however, even this claim is too strong. At most, the *justification* of our set-theoretic beliefs is explained in the way that the justification of our beliefs about gluons is explained. Quine does not show that the reliability of our set-theoretic beliefs is explained in this way. The observable data would have been different if gluons were free, and our beliefs would have varied accordingly. How would the data have varied if there had been no inductive set? This is less clear![30] As with Russell's epistemology, Quine's at most addresses the *justificatory* challenge, not the reliability challenge.

---

[29] Tarski appears to have harbored a similar view. See White [1987].

[30] As are the answers to the contrapositives of such counterfactuals. This will turn out to be important in Chapter 4.

However, Quine's epistemology does not even answer the justificatory challenge, if this requires explaining the justification of our belief in *all* of standard mathematics. Quine writes: "I recognize indenumerable infinities only because they are forced on me by the simplest known systematizations of more welcome matters. Magnitudes in excess of such demands, e.g., $\beth_\omega$ or inaccessible numbers, I look upon only as mathematical recreation and without ontological rights" [1986, 400]. The existence of $\beth_\omega$ is already provable in standard set theory, *ZFC*, using the Axiom of Replacement (Section 1.2). Indeed, depending on what one takes $\beth_\omega$ to *be*, its existence is provable in *ZF* minus *AC*. The cardinal numbers can be defined in an alternative way without *AC*. Moreover, while the existence of *IC* is independent of standard set theory, *ZF* (if *ZF* is consistent), we saw that *ZF* + *IC* proves Con(*ZF*), which *ZF* can only prove if it is inconsistent (Section 1.2). Consequently, it is sometimes maintained that belief in an "inaccessible number" is part and parcel to belief in standard mathematics as well.[31]

What is true is that some fragment of *ZFC* is needed to axiomatize our physical theories, as they are standardly formulated. Those theories are the local gauge theories of the standard model and general relativity. They involve analysis, differential geometry, and algebra, all of which presuppose axioms of set theory.[32] Take, for instance, a claim about the intersection of a family of sets of points on a manifold, $\cap x$. In order to prove that $\cap x$ exists, we must form $\{y : y \in \cup x \,\&\, (z)(z \in x \to y \in z)\}$. This requires the *Subsets* (or *Separation*) *Axiom* schema, which says that for any set, z, and any formula, Φ, there is a set containing those members of z that satisfy Φ. That is, $(z)(\exists y)(x)(x \in y \longleftrightarrow (x \in z \,\&\, \Phi))$ (where y is not free in Φ). This is the (apparently!) consistent restriction of (the first-order fragment of) the *Naive Comprehension* schema, $(\exists y)(x)(x \in y \longleftrightarrow \Phi))$ (where y is not free in Φ). The existence of the union of the sets in x, $\cup x$, is itself guaranteed by the *Union Axiom*, that for any set, z, there is a set, $\cup z$, containing the members of members of z. In symbols: $(z)(\exists y)(x)(x \in y \longleftrightarrow (\exists w)(w \in z \,\&\, x \in w))$. It remains to define the points that get collected in the first place. These will be ordered pairs, understood as sets, ultimately constructed out of natural numbers (whose existence is given by the Axiom of Infinity), by way of repeated applications of Subsets and the Powerset Axiom, that, for any set, z, there is a set containing the

---

[31] Again, this is debatable because *ZF* + *IC* is stronger than *ZF* + Con(*ZF*), which is stronger than *ZF*.

[32] According to these theories, charged objects curve (external or internal) spaces and are affected by the curvature in turn. Potentials are connections giving the curvature of the space. Just as we are free to choose coordinates in general relativity (gravity), we are free to choose a phase in charge space (electromagnetic force), which axes to call the electron and neutrino axes in isospin space (weak force), and which axes to call red, green, and blue in color space (strong force). The objective facts in all cases are those that are indifferent to our local "coordinates," that is, gauges.

subsets of $z, P(z)$, written $(z)(\exists y)(x)(x \in y \longleftrightarrow (w)(w \in x \rightarrow w \in z))$. So several set-theoretic axioms are already implicated in rudimentary claims of physics.

2.2 Nominalistic Science

It might, however, be doubted that we should believe our physical theories *as they are standardly formulated*. Perhaps those theories afford a false but convenient shorthand for better theories that do not speak of mathematical entities. If that were so, then Quine could not explain the justification of our belief in *any* mathematics, as discussed. Field writes, "*even on the assumption that mathematical entities exist*, there is a *prima facie* oddity in thinking that they enter crucially into explanations of what is going on in the non-platonic realm of matter .... [T]he role of mathematical entities, in our explanations of the physical world, is very different from the role of physical entities in the same explanations ... [because f]or the most part, the role of physical entities ... is causal: they are assumed to be causal agents with a causal role in producing the phenomena to be explained" [1989, 18–19, italics in original]. Indeed, this difference might seem to be reflected in scientists' attitudes toward mathematical postulates. Maddy notes that "physicists seem happy to use any mathematics that is convenient and effective, without concern for the ... existence assumptions involved" [1997, 15]. By contrast, the postulation of new fields (particles) is scrutinized. Theories of quantum gravity appear to betray scientists' instrumentalist attitude toward mathematics. These theories commonly postulate that space-time is discrete at the Planck scale but nevertheless use continuous variables (Hagar 2014, Ch. 7).[33]

The standard reason for favoring nonmathematical – that is, *nominalistic* – surrogates to our physical theories as they are currently formulated is that the former avoid the reliability challenge (Field [1989, Intro. 4.A]).[34] But Field also suggests that nominalistic theories afford "intrinsic" explanations, and intrinsicality is a theoretical virtue [1980, 44–45]. Field's use of "intrinsic" is elusive. Prima facie, "intrinsic" cannot just mean causal, contra Field's reference to causality. Had arithmetic been inconsistent, a computer checking for would have said

---

[33] Another incongruence is that quantum theories of gravity incorporating discrete space-time should at least be *consistent* with the claim that the universe is finite. But even if a theory itself admits of finite models, its metatheory, as ordinarily understood, will not. It will be bi-interpretable with arithmetic (which has only infinite models). So an authentic physical theory according to which the universe might be finite needs an "ultrafinitistic" surrogate for the theory of syntax, as well as set theory. More on the problem of metalogic in Section 3.2.

[34] Again, there is also the justificatory challenge for realism about higher (scientifically inapplicable) set theory. But this would be moot if the reliability challenge appeared unanswerable (Section 1.4).

so. So, by ordinary standards, at least some mathematical explanations *do* seem to be causal.[35] Perhaps "intrinsic" means local? If physical facts depended on mathematical ones, then this would involve objects "operat [ing] upon and affect[ing] other matter without mutual contact" [Newton 2007]. The problem with this is that *Bell's Theorem* is widely taken to show that *any* formulation of quantum mechanics – and, hence, of physics generally – must be nonlocal (as the Copenhagen, de Broglie-Bohm, and *GRW* formulations are).[36] Nor would it help invoke the abstract/concrete distinction. Again, if "abstract" means noncausal, then the proposal is at best highly suspect. If it means lacking space-time location, then fundamental particles may turn out to be abstract! Already in nonrelativistic quantum mechanics, particles cannot be assigned trajectories through space-time. But, in quantum field theory, the situation is much more dramatic. There is not even a position operator for photons, for example.[37] Indeed, the *Reeh-Schlieder Theorem* limits talk of localized states in quantum field theory very generally.[38] Chen [2019] takes intrinsicality to be a matter of nonarbitrariness. But some arbitrariness seems unavoidable. For instance, any regimented theory will have to choose a set of logical connectives to take as basic. But it is hard to imagine a principled reason to take, say, $(\exists x)$ as basic and (x) as defined, rather than the other way around (or rather than taking both as basic). Chen's reformulation of (part of) nonrelativistic quantum mechanics betrays additional kinds of arbitrariness, as Chen notes.[39] The most that we can hope for is that "which features of the model are genuinely representational and which are artifacts" is discernible [Sider 2021, 4].

Avoiding reference to mathematical entities is trivial if we are allowed to use certain tricks. For example, if we help ourselves to the operator, *it is mathematically*

---

[35] Recall that Con(*PA*) is a mathematical conjecture that is consistent to deny, if Con(*PA*) (Section 1.1).

[36] This is not beyond dispute because Bell's Theorem assumes that measurements have a unique outcome (contra Everett's interpretation), that there is not a global experimental conspiracy (contra "superdeterminism"), and that that measurements do not affect the prior states of the particles that are measured (contra "retrocausality").

[37] For details, see https://physics.stackexchange.com/questions/492711/whats-the-physical-mean-ing-of-the-statement-that-photons-dont-have-positions

[38] Of course, if any theory of quantum gravity (marrying quantum field theory to general relativity) is true according to which space-time is a manifestation of non-spatiotemporal building blocks, then "abstract" had better not mean lacking space-time location a fortiori. Otherwise, the fundamental ingredients of physical reality would be abstract! Rovelli, a vocal advocate of Loop Quantum Gravity, writes, "quanta of the [gravitational] field cannot live in spacetime; they must 'build' spacetime themselves .... The key conceptual difficulty of quantum gravity is therefore to accept the idea that we can do physics in the absence of the familiar stage of space and time" [2007, 10].

[39] Chen observes, "Instead of invoking a two-place relation Amplitude-greater-than-or-equal-to, whose bearers are pairs of N-regions, we can invoke a 2N-place relation that obey the same axioms but whose bearers are points in Newtonian space-time" (where N is the number of particles in the universe; personal correspondence, reprinted with Chen's permission.)

necessary that P, and take this operator as a logical primitive, then we can believe every *sentence* of pure mathematics without believing in mathematical entities (although mixed, mathematical and nonmathematical, statements will present problems).[40] Alternatively, if we simply assume that the physical world has sufficient structure, then we can find models of our theories in it. Take "1" to refer to the left half of my desk, "2" to refer to the left half of the left half, "3" to refer to the left half of the left half of the left half, and so on to get a model of *PA*! Finally, *Craig's Theorem* ensures that for any first-order theory including mathematical language, there exists a recursively axiomatized nonmathematical theory with the same nonmathematical consequences (Putnam [1965]).

In order to produce *attractive* nominalistic rivals to our best scientific theories, one would need to write down, among other things, a nominalistic "surrogate" for the Standard Model's Lagrangian (density). It is hard to envision what this would look like. But Field [1980] has suggested an approach to nominalizing science, applicable at least to a (substantivalist interpretation of) classical gravitation theory. His reasoning parallels Hilbert's in connection with infinitary mathematics.[41] If N_1, N_2 … are "nominalistic" premises (i.e., they do not quantify over mathematical entities), M is a mathematical theory, and C is a consequence of N_1, N_2 … $+ M$, then Field maintains that C is a consequence of N_1, N_2 … on their own.[42] Mathematics is *conservative* over nonmathematical science. (It follows that mathematics is consistent. If it were inconsistent, then everything would be a consequence of it.)[43] What notion of consequence does Field have in mind? If it is first-order consequence, then mathematics is *not*

---

[40] See Putnam [1967] for a view along these lines.

[41] Hilbert sketches the shape of his program in his 1983/1936 work.

[42] Technically, since N_1, N_2 … might rule out the existence of mathematical entities, for any nominalistic claim, N, if N^* is N with its quantifiers restricted to nonmathematical entities, then Field's contention is that N^* is not a consequence of $N^* + M +$ "[T]here are non-mathematical entities" if N is not a consequence of N_1, N_2 … alone. This does assume a principled distinction between mathematical and nonmathematical entities, something that we have seen that one could doubt. But Field might justify this distinction in terms of the success of the resulting theory.

[43] To see why one might think this, consider applied arithmetic. Using it, we may infer that "we have five apples" from "I have two apples" and "Jenn has three more." But arithmetic is not necessary to underwrite inferences about apples. We could have reasoned using only first-order logic with identity. We do not because that reasoning is unnecessarily complex. We would have to argue:

$(\exists x)(\exists y)[Ax \,\&\, Ay \,\&\, Hix \,\&\, Hiy \,\&\, x \neq y \,\&\, (z)[(Az \,\&\, Hiz) \rightarrow z = x \lor z = y]]$

$(\exists x)(\exists y)(\exists z)[Ax \,\&\, Ay \,\&\, Az \,\&\, Hjx \,\&\, Hjy \,\&\, Hjz \,\&\, x \neq y \,\&\, x \neq z \,\&\, y \neq z)$
$\&\, (q)[(Aq \,\&\, Hiq) \rightarrow q = x \lor q = y \lor q = z]]$

Hence,

$(\exists x)(\exists y)(\exists z)(\exists q)(\exists r)[Ax \,\&\, Ay \,\&\, Az \,\&\, Aq \,\&\, Ar \,\&\, x \neq y \,\&\, x \neq z \,\&\, y \neq z \,\&\, x \neq q \,\&\, x \neq r \,\&\, y \neq q \,\&\, y \neq r \,\&\, z \neq q \,\&\, z \neq r \,\&\, q \neq r \,\&\, (Hix \lor Hjx) \,\&\, (Hiy \lor Hjy) \,\&\, (Hiz \lor Hjz) \,\&\, (Hiq \lor Hjq) \,\&\, (Hir \lor Hjr)]$

conservative, by Gödel's theorems (Shapiro [1983]). If $N_1, N_2 \ldots$ is a consistent and recursive nominalistic theory of space (or space-time) interpreting Robinson Arithmetic, for example, and M is a sufficiently stronger theory, then $N_1, N_2 \ldots + M$ will prove Con $(N_1, N_2 \ldots)$, which $N_1, N_2 \ldots$ cannot prove. On the other hand, if Field has in mind a second-order semantic notion, then conservativeness is not about what we can derive. It is about what is true in all (full) models, with unclear relevance to nominalism.[44]

Notwithstanding the difficulties facing his approach, Field's program is actively pursued. Arntzenius and Dorr [2012] sketch a version of relativistic gravitation theory that does not quantify over "mathematical entities," and Balaguer [1996] (like Chen [2019]) outlines a nominalistic version of (components of) nonrelativistic quantum mechanics.[45] Nevertheless, as of this writing, no one, to my knowledge, has even gestured at a nominalist surrogate for the Standard Model, much less a potential Grand Unified Theory or Theory of Everything, like String Theory.

2.3 The Problem of Metalogic

Successful nominalization requires proof of metatheorems, like conservativeness. Moreover, *everyone* – realist or nominalist – needs to be able to talk about what follows from what and what does *not* follow from what (e.g., a contradiction). The initial metalogical problem for nominalism is that as standardly understood, this is tantamount to talk of proofs or models, neither of which would be familiar physical things.

It is tempting to think that proofs, at least, are more epistemically innocent than numbers. But this is confused. Proofs are made of symbol *types*, universals, whose instances are concrete marks.[46] Hilbert [1983/1936] obscures this when he writes, "In number theory we have the numerical symbols *1, 11, 111, 1111* where each numerical symbols intuitively recognizable by the fact it contains only *1*'s. ... *3 > 2* ... communicate[s] the fact that ... the ... symbol [*2*] is a proper part of the [symbol *3*]" [1983/1926, 193]. Hilbert must have in mind symbol types rather than tokens because, otherwise, it would be a doubtful empirical conjecture that every natural number has a successor. Indeed, the same argument shows that he must understand types *platonistically*, that is, as existing independent of their instances. But it is far from evident that *2* is a proper part of *3*, for it is unclear whether types have parts. In general, platonic types are no more perceivable than numbers. They do not have shape. So it is even specious to suggest that a proof is "surveyable," much less "a concrete object ... a finite configuration ... of

---

[44] See Bueno [2020, Sec. 3] for discussion.

[45] The scare quotes register the obscurity of the mathematical/nonmathematical distinction alluded to previously.

[46] See Wetzel [1989].

recognizable symbols" [Huber-Dyson 1991, 16]. Given their nature, "The relations of dependence between ... the axioms and the theorems" *cannot* "be fully 'visible': their properties and features ... read off from the purely syntactic and structural connections between (the shapes of) the strings" [Berto 2009, 41].[47]

It is true that sometimes, one can trade talk of types for talk of their instances. Instead of saying "the letter O is oval in shape," one can say "letter Os are oval in shape." But this will not work for uninstanced types, since the corresponding general claims are vacuously true. For instance, "proofs of length at least 10^{10} are finite" is no more true than "proofs of length at least 10^{10} are infinite," assuming no such proof has been instanced. We might trade claims about types for *necessitated* general claims, such as "necessarily, for any, x, if x is a proof, then x is finite." However, explaining the reliability of our belief in such modal claims seems no easier than explaining the reliability of our belief in metalogical claims (see Section 4.2).

In order to address this problem, Field develops a nominalist metalogic ([1989, Introduction, Part II] and [1991]). He argues that consistency is a theoretical primitive, like "and" or "not." We appeal to judgments of what is consistent in deciding what models and proofs there are, rather than the other way around. He writes, "there are 'procedural rules' governing the use of ['consistent' and 'implies'], and ... these ... give the terms the meaning they have, not ... definitions, whether in terms of models or of proofs" [1991, 5]. But this argument is too quick. Neither a model- nor a proof-theoretic reductionist should claim that our *knowledge* of what is consistent depends on our knowledge of what models or proofs there are. By that reasoning, Lewis [1986] is not a reductionist about metaphysical possibility, since he appeals to (epistemically) prior judgments about what is possible in deciding what concrete worlds there are. The real question is whether the avoidance of Field's ideology is worth the mathematical ontology. This is not obvious.

Field requires two sentential operators, $<>$ and $[]$. These are read "it is logically possible that" and "it is logically necessary that," respectively. They are dual in the standard sense, so $[]P \longleftrightarrow \sim <> \sim P$ and $<> P \longleftrightarrow \sim[]\sim P$. Field also invokes a substitutional quantifier, # Φ, as a device for infinite conjunction. This allows him to assert the infinitely many axioms of *ZFC* (again, some of these are given by schemas), rather than saying *of* them that

---

[47] Bourbaki [1970, Chapter 1] also seems not to appreciate the distinction between types and tokens. Later, however, Chavelley, a leading figure of the Bourbaki group, appears to recognize the problem. He writes, "the idea of a symbol which is 'the same,' although written in different places and at different times, is not at all an idea that stands by itself. But it must stand by itself if one has this conception ... of mathematics. Not only can this idea not possibly be realized, but its content is absurd. A symbol cannot possibly be 'the same' if it does not have an aura of signification. There ... is an appeal to something human that contradicts the idea of a perfect 'horizon' [i.e., complete rigor]" [Guedj, 1985].

they are true (which would be to speak of mathematical entities – namely, symbol types).[48] For example, all instances of the Subsets Axiom schema would get expressed as follows: $\# \Phi(z)(\exists y)(x)(x \in y \longleftrightarrow (x \in z \,\&\, \Phi))$.

On Field's account, if AX_T is the conjunction of the axioms of a mathematical theory, T, then what we *really* know is that $<> AX_T$ and that $[](AX_T \to P)$, for "proved" theorems, P. Thus, a difficulty facing this position is to explain why reasoning about models and proofs is so useful in metalogic, given that it is not really about proofs or models. We infer that $<> AX_T$ from the premise that AX_T has a model. And we infer that $\sim <> AX_T$ from the premise there is a proof of $(P \,\&\, {\sim}P)$ from AX_T. How can Field explain this? He makes an argument inspired by Kreisel's "squeezing" argument (Kreisel [1967b]). Using Field's symbolism, the squeezing argument is as follows:

1. If $[](P \to Q)$, then every model of P is a model of Q.
2. If there is a proof from P to Q (in some standard first-order proof system), then $[](P \to Q)$.
3. *Gödel's Completeness Theorem*: If every model of P is a model of Q, then there is a proof from P to Q.
4. So, $[](P \to Q)$ just in case there is a proof of Q from P, just in case every model of P is a model of Q. (Logical necessity is *coextensive* with truth in all models and derivability.)

Field cannot accept this argument, however. According to him, we do not know that (1)–(3) are (non-vacuously) true! In order to explain the usefulness of proofs and models, Field reasons as follows:

1. If $[](P \to Q)$, then $[](AX_{ZFC} \to$ [every model of P is a model of Q]).
2. If $[](AX_{ZFC} \to$ [There is a proof from P to Q]), then $[](P \to Q)$.
3. $[](AX_{ZFC} \to$ ([every model of P is a model of Q] $\to$ [There is a proof from P to Q])).
4. So, given that $<> AX_{ZFC}$, $[](P \to Q)$ just in case $[](AX_{ZFC} \to$ [every model of P is a model of Q]), just in case $[](AX_{ZFC} \to$ [There is a proof from P to Q]).

The problem with Field's position is that once we give up on intrinsicality, we are left with the reliability challenge as the motivation for nominalism.[49] But the reliability of our belief in, say, $<>(ZFC)$ does not appear to be any easier to explain

---

[48] Field actually focuses on the finitely axiomatizable theory, *NBG*. But he needs a device for infinite conjunction in any event. So I continue to speak of *ZFC* because it has already been introduced.

[49] The justificatory challenge for realism about specifically scientifically inapplicable mathematics might also seem to favor nominalism. But, in Section 3.5, we will see that the justificatory (and reliability) challenge for mathematical realism can arguably be reduced to that for arithmetic realism.

than that of our belief in Con(*ZFC*). Putnam [2012] suggests that the latter concerns abstract objects, while the former does not. But the reliability challenge does not depend on an ontologically committed interpretation of our mathematical theories. (If it did, then moral realists could avoid such a challenge simply by adopting nominalism about universals (Clarke-Doane [2020b, Section 1.5]). We could even state the challenge so as to avoid reference to truths. The problem is to explain mathematical instances of the schema: in general, if mathematicians believe *P*, then *P* (where we *use*, and do not mention, *P*).

Whether our concern is with $<>$ (*ZFC*) or Con (*ZFC*), our failure to derive a contradiction in *ZFC* does not illuminate much (contra Field [1989, 89]). As Leng writes, "Unless we have reason to believe that the derivations we are able to produce so far are a suitably representative sample of all possible derivations, this kind of enumerative induction will provide only a very weak justification for our belief" [2007, 105]. In particular, if "any contradiction in our [mathematical physics is] only derivable in a derivation too long for humans to produce, then the best explanation of the applicability of that piece of mathematics might [not] require that it is consistent" [2007, 106]. So the reliability (and, indeed, justificatory) challenge for realism about consistency –which is tantamount to arithmetic realism –is acute for nominalists too.

2.4 Conclusion

I have discussed Quine's response to the justificatory challenge and Field's nominalist rejoinder. Even if some of our mathematical beliefs are justified empirically, in the way that our beliefs about electrons or gluons are justified, it is hopeless to argue that all of our mathematical beliefs are justified in this way. Field argues that *none* of our mathematical beliefs is justified empirically because our current scientific theories approximate better theories that make no reference to mathematical entities. However, attractive "nominalistic" surrogates for them are lacking, and it is unobvious what could recommend them even if they existed. Mathematical entities are causal by ordinary criteria. Conversely, nominalist physical theories incorporate formal artifacts, are generally nonlocal, and postulate objects that lack space-time locations. Finally, nominalist theories must at least assume arithmetic, or its modal surrogate, at the level of metalogic. Thus, the reliability challenge arises even for nominalists. Let us turn to that now.

3 Connection, Contingency, and Pluralism

I have discussed the justificatory challenge, that is, the challenge to explain the defeasible justification of our mathematical beliefs, and responses to it in terms of self-evidence, analyticity, reflective equilibrium, and scientific application. It remains to consider the reliability challenge. What explains the non-coinciden-tal truth, that is, reliability, of our mathematical beliefs? In this section, I substantially clarify this challenge and outline the most promising response to it, *mathematical pluralism*. I conclude with pluralism's revisionary metatheore-tic implications.

3.1 Clarifying the Challenge

This Element began by segregating the challenge to explain our knowledge of mathematics into two aspects, corresponding to the two components of know-ledge. Those aspects are the justificatory and reliability challenges. Although I argued that the former is acute, the latter has been the focus of discussion since Benacerraf [1973]. In that article, Benacerraf insists that "something must be said to bridge the chasm, created by ... [a] ... realistic ... interpretation of mathematical propositions ... and the human knower" [1973, 675]. Absent such an account, "the connection between the truth conditions for the statements of [our mathematical theories] and ... the people who are supposed to have mathematical knowledge cannot be made out" [1973, 673].

The request for an "account of the link between our cognitive faculties and the objects known" is widely interpreted as the reliability challenge. Consider Gödel's remark that "despite their remoteness from sense experience, we ... have something like a perception also of the objects of set theory as is seen from the fact that the axioms force themselves upon us as being true" [1947/1983, 483–4]. He complains, "I don't see any reason why we should have less confidence in this kind of perception, that is, mathematical intuition, than in sense perception" [1947/1983, 483–4]. Gödel's remarks are responsive to the justificatory chal-lenge. Appeals to intuition help explain the (defeasible) justification of our set-theoretic beliefs, just as appeals to observation help explain the justification of our perceptual beliefs (Section 1.4). Perhaps it "seems" to us that for any two sets, there is their pair set, in something like the way that it seems to us that here is a hand. However, Gödel's remarks are unresponsive to Benacerraf's challenge. Why would being the content of an intuition be a reliable symptom of being true? Benacerraf emphasizes, "In physical science we have at least a start on such an account, and is causal" [1973, 674]. But no causal account suggests itself in the case of (pure) mathematics. Certainly, sets do not deflect photons that hit our retinas, stimulating our optic nerves![50]

---

[50] This is so even if sets may be causally efficacious, as argued in Section 2.2. See, however, Van Atten and Kennedy [2009] for reasons to think that Gödel would not have accepted the dialectic as Benacerraf conceives of it.

Notwithstanding common allegations to the contrary (see, e.g., Hart [1996], Colyvan [2007, 111]), Quine's empiricist epistemology is also vulnerable to Benacerraf's criticism. The fact that truths about sets are implied by the best explanation of our observations does nothing, by itself, to show that our observations are responsive to those truths in the way that they are responsive to truths about gluons, for example. Indeed, every logical truth is implied by every explanation at all. But it is not that easy to explain the reliability of our logical beliefs! As stressed in Section 2.1, the reliability challenge is pressing for empiricists as well (Field [1989, 22–3]).

Although Benacerraf first drew attention to a challenge in the vicinity, the canonical presentation of the reliability challenge is actually due to Field (Liggins [2010]; Linnebo [2006]). He writes:

> We start out by assuming the existence of mathematical entities that obey the standard mathematical theories; we grant also that there may be positive reasons for believing in those entities … .But Benacerraf 's challenge … is to … explain how our beliefs about these remote entities can so well reflect the facts about them … .*[I]f it appears in principle impossible to explain this*, then that tends to *undermine* the belief in mathematical entities, *despite* whatever reason we might have for believing in them. [Field 1989, 26, emphasis in original]

So formulated the the reliability challenge is of considerable interest. First, it cannot be dismissed as a puzzle of no practical import. If sound, the apparent impossibility of answering the challenge *undermines* our mathematical beliefs, realistically construed.[51] So, absent an answer to it, we ought to *change our beliefs*. The challenge might not have this upshot if it just showed, as Benacerraf suggests, that our mathematical beliefs fail to qualify as knowledge. If our beliefs are justified and we can explain their reliability, then who cares if they qualify as knowledge?

Second, Field's formulation is not a generic "convince the skeptic" challenge. He *grants* that our mathematical beliefs are (actually) true and (defeasibly) justified (under a realist construal). He argues that it appears impossible to explain their reliability *given these assumptions*. If Field did not grant these things, then his challenge would overgeneralize. The evolutionary explanations of our having reliable mechanisms for perceptual belief, and the neuro-physical explanations of how those mechanisms work such that they are reliable, all

---

[51] The difference between the conclusion that it undermines our mathematical beliefs and the one that it undermines our belief in mathematical objects, but not our mathematical beliefs, is merely one of natural language semantics. The nonverbal claim is that the impossibility of answering the challenge undermines our belief that, for example, there are infinitely many prime numbers, *as the realist interprets this*, whether or not realists are right about natural language. This is why I add the caveat "realistically construed." See the Introduction for the prima facie case for realism.

presuppose the (actual) truth of our perceptual beliefs (under a realist construal).[52] These explanations do not *state* that our perceptual beliefs are reliable. But if we were not justified in supposing that our perceptual beliefs were reliable, then we would not be justified in accepting the perceptual evidence to which those explanations appealed. Field's contention is that there is a relevant *difference* between our mathematical and perceptual beliefs.

Third, Field's challenge is distinct from the challenge to explain the *determinacy* of our mathematical beliefs (Field [1989, Introduction, Part I]; Putnam [1980]).[53] In fact, the challenges are in tension. According to the challenge to explain the determinacy of our mathematical beliefs, there is nothing in our practice or the world that could pin down an "intended model" of set theory.[54] For instance, if *ZFC* is consistent, then so is *ZFC* + *CH* and *ZFC* + ~*CH*.[55] So what could make it the case that we are talking about a (class) model in which *CH* is determinately true or false ([Martin 1976, 90–1])? There are no apparent physical relations between us and sets that could help to explain determinate reference (Field [1998]; Putnam [1980]). One could always appeal to "natural kinds" à la Lewis [1983]. But, absent an account of what *makes* one model more natural than another, and of how naturalness facilitates reference, this response is without much content. Some philosophers conclude that if a mathematical statement is undecidable with respect to the (first-order) axioms that we accept then it must be indeterminate. Note, however, that the fewer (determinate) truths one postulates, then the fewer determinate truths one must explain our reliability with respect to. So the reliability challenge arises *only to the extent that* the determinacy challenge can be answered.

The final point is that despite Benacerraf's and Field's talk of objects, the reliability challenge does not really depend on an ontologically committal interpretation of mathematics. It depends on mathematical realism – that is, the view that there are (non-vacuous) mathematical truths that obtain independent of minds and languages.[56] It does not matter whether those truths owe themselves to the existence of special entities, like sets. To be sure, various

---

[52] See Schechter [2010] for the distinction between these two levels of explanation.

[53] Barton [2016] seems to conflate these challenges, as does Benacerraf himself in his [1973].

[54] I put "intended model" in quotation marks because, in the case of set theory, it is a proper class, and so not strictly a (set) model.

[55] We saw that Con $(ZFC) \rightarrow Con(ZFC + CH)$ in Section 1. To show that $Con(ZFC) \rightarrow Con(ZFC + \sim CH)$, one constructs a *generic extension*, $M[G]$, of a countable transitive model, M, of ZFC and adds κ^M – many new subsets of ω to M, without collapsing any cardinals, resulting in a "wider" model of the same height. This ensures that $M[G]$ acquired $\kappa = \aleph_k$ new subsets (according to the model), for our choice of k so that $M[G] \mid = 2^{\aleph_0} = \kappa$. See Cohen [1966].

[56] This is what Shapiro calls "realism in truth-value" in his [1997, 37] work.

realists have suggested that their preferred version of realism faces no reliability challenge because it is ontologically innocent (Putnam [2012]; Scanlon [2014, 70]; perhaps Tait [1986]). But this is confused. If that were correct, then the reliability challenge would have no application to moral realism, assuming the nominalism about properties of Quine [1948], or to Field's realism about primitive logical possibility (Section 2.3). The challenge is to explain mathematical instances of the schema: if mathematicians believe that *P*, then *P*. This arises whether or not *P* happens to carry with it any ontological commitments.

However, while Field's formulation of Benacerraf's reliability challenge is compelling, it is unclear at a crucial juncture. It is unclear what it would take to *explain the reliability* of our beliefs of a kind, *F*, in the requisite sense. That sense is one according to which (a) it appears impossible to explain the reliability of our mathematical beliefs (realistically construed) and (b) if it appears impossible to explain the reliability of our belief that *P*, then this *undermines* that belief (so construed). (An *undermining*, as opposed to *rebutting*, defeater of our belief that *P* is a reason to lower our credence in *P* that is not also a *direct* reason to believe that ~*P*.) It is not transparent what, if any, interpretation of "explain the reliability" might satisfy both (a) and (b).

3.2 Connection

The standard interpretation of "explain the reliability" originates with Benacerraf [1973]. He writes, "the connection between the truth conditions for statements of number theory and any relevant events connected with the people who are supposed to have mathematical knowledge cannot be made out" [1973, 671–3]. The mathematical truths are "up there," and we are "down here," and there is no glue linking the two. In the present context, the suggestion is as follows:

> *Answer 1 (Connection)*: In order to explain the reliability of our beliefs of a kind, *F*, it is necessary to show that for any one of them, *P*, our (token) belief that *P* is *connected* to the fact that *P*.

What kind of connection is in question? The relevant kind could be causal (Benacerraf [1973, 671–673]; Cheyne [1998]; Goldman [1967]), noncausal but explanatory (Faraci [2019]), or logical (Joyce [2008, 2016]). However, none of these interpretations makes both (a) and (b) true. Consider (a) first. We saw in Section 2.3 that by ordinary standards, there *is* a causal connection between our mathematical beliefs and the mathematical facts. At least some of the former counterfactually depend on the latter.[57] Benacerraf might reply that this is not

---

[57] See Montero [2019] for another defense of the view that mathematical facts are causally efficacious.

the right kind of connection. But, absent a non-circular characterization of that kind, this rejoinder is without force. Similarly, the idea that mathematical facts explain – even if they do not cause – our (token) mathematical beliefs is commonplace. Consider the suggestion that "[p]eople evolved to believe that $2 + 3 = 5$, because they would not have survived if they had believed that $2 + 3 = 4$" [Sinnot-Armstrong 2006, 46]. Why would they not have survived? The obvious answer is that "they would not have survived [because] it is true that $2 + 3 = 5$" [Sinnot-Armstrong.].[58] Finally, there is a trivial logical connection between at least many of our (token) mathematical beliefs and the facts, if mathematics is indispensable to empirical science in the way that it seems to be. Steiner notes, "suppose that we believe ... the axioms of analysis or of number theory .... [S]omething is ... responsible for our belief, and there exists a theory ... which can satisfactorily explain our belief .... *This theory, like all others, will contain the axioms of number theory and analysis*" [1973, 61, italics in original].

However, even if (a) were true under any of the previous interpretations, (b) would be false. The causal theory of knowledge has been widely rejected for reasons that have nothing to do with the reliability challenge (Field [2005, 77]). But if it is implausible that knowledge that P requires a causal connection to obtain between our belief that P and the fact that P, then it is implausible that justified belief that P requires the appearance that this is so. What of *Answer 1* under the noncausal explanatory or logical readings of "connection"? The problem with these is that underminers (as opposed to rebutters) must arguably be modal (Baras & Clarke-Doane [2021]; Clarke-Doane [2015]). If evidence neither tells directly against our belief that P (and so is not rebutting) nor suggests a mismatch between our belief and the truth in (what are perhaps) other circumstances, then why give it up? The impossibility of showing that our beliefs are explanatorily or logically connected to the truths does nothing to suggest this kind of mismatch.

Suppose, for instance, that in addition to being justified in thinking that our mathematical beliefs are actually true (something Field grants), we are justified in believing that the mathematical truths (whatever they are) are necessary and that our beliefs could not have been different in any relevant circumstance (because, say, they are evolutionarily innate). Then we would be justified in thinking that our beliefs could not have been false in any relevant circumstance, even if they were not explanatorily or logically connected to the facts. Of course, the mathematical truths may not be necessary, and our mathematical beliefs may not be innate (I will argue as much shortly). The point is that

---

[58] Sinnott-Armstrong does not say explicitly that he takes this to be a noncausal explanation. But he nowhere indicates sympathy for the heretical view (defended here) that mathematical entities are causal. (To be clear, I am *not* saying that the obvious answer is true [see Clarke-Doane (2012) for the contrary view]. I am saying that it does not "appear in principle impossible" to show that it is.)

evidence that there is no explanatory or logical connection between our beliefs and the facts is no reason to think so.

3.3 Counterfactual Dependence

These considerations suggest that lack of connection only matters insofar as it is indicative of "modal insecurity" (Clarke-Doane [2015]). There are two cases to consider: situations in which the truths are different but our beliefs fail to be and situations in which our beliefs are different but the truths fail to be.

Field often takes explaining the reliability of our mathematical beliefs to involve showing that "if the ... facts had been different then our ... beliefs would have been different too" [1996, 371]. In particular, "The Benacerraf problem ... seems to arise from the thought that we would have had exactly the same mathematical ... beliefs even if the mathematical ... truths were different ... and this undermines those beliefs" [2005, 81]. He is particularly concerned that "we can assume, with at least some degree of clarity, a world without mathematical objects" [2005, 80–1].[59] In such a world, Field worries, our mathematical beliefs would have failed to vary correspondingly.

As stated, the reliability challenge does not satisfy (b), even if it satisfies (a). Had there been no "perceptual objects" – that is, objects of ordinary perception – then our perceptual beliefs may well have been unaffected too. We may have been dreaming, hallucinating, or deceived by an evil demon. This is just another way of formulating the skeptical import of Descartes's *cogito*. The sensible demand in the neighborhood is to show that our perceptual beliefs are *sensitive*. That is:

> *Answer 2 (Sensitivity)*: In order to explain the reliability of our beliefs of a kind, F, it is necessary to show that for any one of them, P, had it been the case that $\sim P$, we would not still have believed that P (using the method that we used to determine whether P).[60]

Unlike the Field's formulation, this one is not obviously too demanding. Had I not had a hand, the world would have been similar in other respects, and I would not have believed that I had a hand (using the method that I actually used). Had $1+1 \neq 2$, would I still have believed that $1 + 1 = 2$? This is less clear. It might be thought that there is an evolutionary argument that I would not. But the obvious argument trades on an equivocation between arithmetic truths and (first-order) logical truths (Clarke-Doane [2012]). Let us grant that had the

---

[59] Alternatively, under an ontologically innocent interpretation of mathematics, we can imagine a world in which there are no non-vacuous mathematical truths.

[60] The "for any one of them" quantifier is too strong. Perhaps "for most of them" or "for a typical one of them" would serve. This complication will be irrelevant.

following counter-logical been (non-vacuously) true, we would have believed it. If there is "exactly one" lion to the left and "exactly one" lion to the right and no lion to the left is a lion to the right, then there are no lions to the left or to the right (where the phrases "exactly one" and "exactly two" here are abbreviations for constructions out of ordinary quantifiers and the identity sign and do not refer to numbers). That is, we would have believed the following, had it been the case: $[\exists x(Lx \& Ax \& (y)([(Ly \& Ay) \to (x = y)])) \& (\exists x)(Lx \& Bx \& (y)([(Ly \& By) \to (x = y)]) \& \sim \exists x(Lx \& Ax \& Bx)] \to \sim(\exists x)(Lx \& (Ax \vee Bx))$, where "$Lx$" means that x is a lion, "Ax" means that x is to the left, and "Bx" means that x is to the right. Still, what we actually need to establish is that had the number *1* borne the plus relation to itself and to *0* and *the (first-order) logical truth* that if there is exactly one lion to the left and exactly one lion to the right and no lion to the left is a lion to the right, then there are exactly two lions to the left or to the right *held fixed*, then we would not have believed that *1 + 1 = 2*. The closest worlds in which the mathematical truths are different are presumably still worlds in which the logical truths are the same (not to be confused with worlds in which the *meta*logical truths, that such truths as the aforementioned *are* logical truths, are the same). But this is doubtful. Had the number *1* borne the plus relation to itself and to *0*, but *it remained true* that if there is "exactly one" lion to the left and "exactly one" lion to the right and no lion to the left is a lion to the right, then there are "exactly two" lions to left or to the right, it would have benefited us to believe that *1 + 1 = 2* (see Clarke-Doane [2012, Sec.III])!

However, while (a) is plausible under this reading, (b) is not. The problem is that *our belief in pretty much every "counterpossible" is insensitive if it is not just vacuously sensitive.*[61] For example, had the bridge laws that link subvenient properties to supervenient properties been false, we still would have believed them (using the method that we actually used). It might be thought that the upshot of this is just skepticism about "necessary" truths.[62] But skepticism about necessary truths engenders skepticism about contingent ones. If I believe that I am looking at a piece of paper, but my belief in the necessary bridge law that field excitations arranged "paper-wise" compose a piece of paper is undermined, then it is hard to see how my belief that I am looking at a piece of paper could fail to be. This assumes that justification is closed under known entailment. But how could

---

[61] The qualifier "pretty much" because there are apparent exceptions, like the counter-mathematical mentioned in Section 2.2. Of course, if counter-mathematicals were vacuous then *Answer 2* would not get off the ground. But counter-mathematicals have no claim to being counterpossibles in a sense in which counterpossibles might be vacuous. See Clarke-Doane [2019a].

[62] The scare quotes around "counterpossible" and "necessary" indicate that there is *some* sense in which the relevant claims are *not* impossible (see Field [1989, I.5 & I.6], Clarke-Doane [2019b]), even if they are metaphysically so. In particular, for typical such truths, $P, \sim(\sim P[] \to \perp)$ on the ordinary reading of $[] \to$. More on this in Section 4.

beliefs about the conditions under which properties are instanced be rationally insulated from beliefs ascribing them? *Answer 2* is too strong.[63]

3.4 Contingency

For purposes of the reliability challenge, the situation to worry about is, thus, one in which our *beliefs* are different, while the truths fail to be. But the problem cannot just be that had our mathematical beliefs been different, they would have been false. That is another way of saying that the mathematical truths do not counterfactually depend on our beliefs. This is part and parcel to mathematical realism. The real worry in the vicinity is that our mathematical beliefs could have *easily* been different (using the method that we actually used). Given that the mathematical truths could not have easily been different, it seems to follow that our mathematical beliefs could have easily been *false* (using the method that we actually used).[64] This suggests the best answer.

> *Answer 3 (Safety)*: In order to explain the reliability of our beliefs of a kind, *F*, it is necessary to show that for any one of them, *P*, we could not have easily had a false belief as to whether *P* (using the method that we used to determine whether *P*).[65]

Answer 3 is the only answer to the initial question of which I am aware that might satisfy (b). Evidence that we could have easily had a false belief as to whether *P* (using the method that we actually used to determine whether *P*) is a paradigm undermining (as opposed to rebutting) defeater of our belief that *P*. It gives us reason to give up our belief that *P*, but not by giving us direct reason to believe that ~*P*. This is arguably how learning that our beliefs in "necessary" truths, like moral or religious ones, are the products of social forces might "debunk" them. It might give us reason to believe that we could have easily had different, and so false, ones.

Is (a) true? It depends on what "easily" means.[66] But set-theoretic beliefs at least appear to have as strong a claim to being unsafe as do paradigmatic philosophical beliefs that are commonly supposed to be, such as modal,

---

[63] See White [2010, 580–581] for counterexamples to the view that apparent insensitivity is undermining that involve only contingent truths.

[64] "Seems to" because "different" is ambiguous in a way that will emerge in Section 3.5.

[65] Safety must actually be complicated slightly in ways that will not matter for what follows. See Clarke-Doane [2020b, Sec. 4.6] for details.

[66] It does not help to say that we could have easily had false *F*-beliefs just in case the probability that our *F*-beliefs are true is low. This just raises the question of how to understand the relevant sense of "probability." It cannot be objective. If the universe is macroscopically deterministic, near enough, then, given its state before our forming a belief in, say, *AC*, Pr(we believe that AC) $\approx$ 1. So, if Pr(AC) = 1, Pr(we believe that $AC \& AC$) $\approx$ 1, which implies that Pr(our belief in AC is true) $\approx$ 1.

(meta)logical, or normative ones (Section 4). Recall the kinds of disagreements that were surveyed in Section 1. The problem was that unlike paradigmatic scientific disagreements, mathematical disagreements tend to bottom out in contingent intuitions (plausibility judgments). That raised the specter that they fail to count as (defeasibly) justified. But there is another way that disagreement could threaten our mathematical beliefs. Even if mathematical intuitions *defeasibly* justify our belief in standard axioms, perhaps that justification is *undermined* by knowledge of their variability. We observe that our belief in *AC* – a fortiori large cardinals, *Projective Determinacy*, and so on – is not inevitable, even given the same evidence, standards of argument, levels of intelligence, education, sincerity, and attentiveness. Instead, it bottoms out in accidental "differences in … taste" [Jensen 1995, 401]. Even if those with a taste for *AC* happened to set the agenda for set theory – somewhat like those with a taste for the *Necessity of Identity* (Section 4.2) set the agenda for modal metaphysics and those with a taste for classical logic (Section 4.3) set the agenda for (meta)logic – we could have ended up skeptics of the orthodoxy. We could have studied under a different advisor or witnessed an enthralling talk by a renegade. Indeed, the orthodoxy itself is conspicuously contingent. Pudlák writes:

> Imagine that the Axiom of Determinacy [which is inconsistent with AC] had been introduced first, and before the Axiom of Choice was stated the nice consequences of determinacy, such as the measurability of all sets, had been proved. Imagine that then someone would come up with the Axiom of Choice and the paradoxical consequences were proved. Wouldn't the situation now be reversed in … that the Axiom of Determinacy would be "the true axiom," while the Axiom of Choice would be just a bizarre alternative? [2013, 221, emphasis added]

Similarly, Hamkins remarks:

> Imagine … that … the powerset size axiom [(*PSA*) that for any x and y, $|x| < |y|$ implies $2^x < 2^y$] had been considered at the very beginning of set theory … and was subsequently added to the standard list of axioms. In this case, perhaps we would now look upon models of ~*PSA* as strange in some fundamental way, violating a basic intuitive principle of sets concerning the relative sizes of power sets; perhaps our reaction to these models would be like the current reaction some mathematicians (not all) have to models of $ZF + {\sim}AC$ or to models of Aczel's anti-foundation axiom *AFA*, namely, the view that the models may be interesting mathematically and useful for a purpose, but ultimately they violate a basic principle of sets. [2012, 432–3][67]

---

[67] As discussed in Section 1, such examples are the tip of the iceberg. Weyl's *predicativism*, or a still more revisionary program, could have won out, in which case we might have rejected standard theorems of analysis.

So, *if* there is an interpretation of "explain the reliability" such that (a) it appears impossible to explain the reliability of our mathematical beliefs (realistically construed) and (b) if it appears impossible to explain the reliability of our belief that *P*, then this undermines that belief, *then* it is apparently the safety interpretation. It is conceivable that there is simply no interpretation of "explain the reliability" that satisfies both (a) and (b) (Clarke-Doane [2016]). In that case, there would be no reliability challenge meeting all of the constraints that have been placed on it.[68]

3.5 Mathematical Pluralism

Suppose, then, for the sake of argument, that lack of safety is undermining and that we could have easily had different mathematical beliefs. The inference to the conclusion that we could have easily had *false* ones – that is, that our mathematical beliefs are unsafe – depends on an assumption that has gone unexamined. This is that the axioms of our foundational theories, like set theory, are not analogous to the Parallel Postulate of (pure) geometry (Section 1.3). If we jettison this assumption, and suppose, at first pass, that we could not have easily had *inconsistent* mathematical beliefs, then (a) can be challenged. Every consistent foundational theory may be true of its intended subject, independent of minds and languages. Field himself concedes:

> [Some philosophers] (Balaguer [1995)]; Putnam [1980)]; perhaps Carnap [1950a]] (1983) solve the problem by articulating views on which though mathematical objects are mind-independent, any view we had had of them would have been correct .... [T]hese views allow for ... knowledge in mathematics, and unlike more standard Platonist views, they seem to give an intelligible explanation of it. [2005, 78][69]

---

[68] In his 1989 work, Field appears to propose another interpretation, in addition to the three kinds considered here. He writes, "If the intelligibility of ... 'varying the facts' is challenged ... it can easily be dropped without much loss to the problem: there is still the problem of explaining the *actual* correlation between our believing 'p' and its being the case that p" [238, italics in original]. But I do not know what this means. It might mean showing that the correlation holds in nearby worlds, so the actual correlation is not a fluke. In that case, though, we are just back to safety if we do not vary the mathematical facts and safety or sensitivity if we do. Perhaps there is a hyperintensional sense of "explanation" according to which one can request an explanation of the "merely actual correlation" between our beliefs and the truths? Only if it avoided the objections to those discussed in connection with *Answer 1*. Again, even if we cannot explain the merely actual correlation between our mathematical beliefs and the truths, in some hyperintensional sense, we might be able to show that they are sensitive and safe (and objectively probable), realistically construed.

[69] Note that this view does nothing to establish a connection between our (token) beliefs and the truths or to show that they are sensitive. So, it would not answer the reliability challenge if *Answer 1* or *2* were instead correct.

Mathematical pluralism, as I will call this kind of view, requires care. Consider the *proposition* that if t is a disjointed set not containing the empty set, then there exists a subset of ∪t whose intersection with each member of t is a singleton (*AC*). If *AC* is true, then had we believed ~*AC*, our belief would have been false – assuming mathematical realism. This is just to say ~(*AC* & ~*AC*). What *might* be true is that had we accepted the *sentence* which we would ordinarily take to express the negation of *AC* – namely,

$$\sim[(t)[(x)[x \in t \rightarrow (\exists z)(z \in x) \& (y) (y \in t \& y \neq x \rightarrow \sim(\exists z)(z \in x \& z \in y))]$$
$$\rightarrow (\exists u)(x)(x \in t \rightarrow \exists w(v) [v = w \longleftrightarrow (v \in u \& v \in x)])]] - \text{that sentence}$$

would have expressed the negation of a *different proposition* out of our mouths, and this negation is true too.[70]

Consider the Parallel Postulate *sentence*, S_{PP}: "two straight lines intersecting another so as to make less than 180° angle on one side intersect on that side." It is false that had we believed the negation of the proposition that we now use S_{PP} to express, our belief would have been true. That proposition is presumably about Euclidean space, and is true independent of us, given mathematical realism. What is plausible is that had we accepted the *sentence*, $\sim S_{PP}$ ("it is not the case that two straight lines ..."), we would have believed a *different proposition*, and it is true right alongside the original. If P_{PP} is the proposition expressed by S_{PP}, then had we uttered $\sim S_{PP}$, we would have asserted a $\sim P_{PP}$– like proposition – where a $\sim P$-like proposition is the translation of $\sim P$ into a possibly distinct true proposition that shares $\sim P$'s "metaphysical content." Different geometrical spaces sit side by side, one no more real than another.[71] The mathematical pluralist maintains that rather than being a special feature of (pure) geometry, this is the general situation with foundational theories.

How general? As sketched previously, mathematical pluralism is general indeed. It entails that *every (first-order) consistent set theory* has an intended model (Balaguer [1995]; Field [1998]; perhaps Hamkins [2012], Leng [2009]). (This is *not* just the view that every such theory has a set model, which follows from the Completeness Theorem. It is the view that every such theory has an intended class model, of the sort that anti-pluralists, like Gödel or Woodin, take ZFC to have.)[72] But if the thesis that the reliability challenge is answerable is not *itself* like the Parallel Postulate (!), then this formulation is untenable.

Consider any (first-order) consistent set theory, *T*, interpreting *PA*. Then, if *T* is consistent, so is *T* conjoined with a coding of the claim that *T* is not consistent,

---

[70] See Rabin [2007] for a response along these lines to something like this problem as formulated in Restall [2003].

[71] See Field [1998] for the "side by side" language in connection with set-theoretic universes.

[72] For a discussion of pluralism about type theory and its relation to pluralism about set theory, see McCarthy and Clarke-Doane [Forthcoming].

$T + \sim\text{Con}(T)$, by Gödel's Second Incompleteness Theorem (Section 1.1). So, if any consistent foundational theory has an intended model, then so does $T + \sim\text{Con}(T)$. But that is tantamount to the view that whether T is consistent is like whether the Parallel Postulate is true! If this is the pluralist's position, then even if *they* can argue that we could not have easily had false mathematical beliefs, *we* can argue for the intuitively opposite conclusion. Even if our belief in *ZF* is safe because *ZF* is consistent and we could not have easily had inconsistent mathematical beliefs, there is a symmetric argument that our belief in *ZF* is false, so certainly not safe, because *ZF* is inconsistent – or "shinconsistent." There is nothing privileged about the pluralist's notion of (classical) consistency. Related consequences of pluralism so generally formulated include that what *ZF* *is* and even what the language of *ZF* consists in are like the question of whether the Parallel Postulate is true. A theory like $ZF + \sim\text{Con}(ZF)$ that "disagrees" with *ZF* about what counts as finite disagrees too about what a theory and language is.

Where, then, should the pluralist draw the line, if not at (first-order) consistency? Perhaps at arithmetic soundness. A theory is *arithmetically sound* when it only implies true arithmetic sentences. If we replace the consistency requirement with an arithmetic soundness requirement, then only set theories that are right about finiteness, consistency, and syntax will count as true of their intended subjects. ($\text{Con}(PA)$, $\text{Con}(ZF)$, and so on are all Π_1 arithmetic sentences.) This means that the absolute or, as I will say, *objective* mathematical truths are not recursively enumerable, as they would be if the pluriverse witnessed every consistent theory. But pluralism might still answer to the reliability challenge, understood as the challenge of showing that our beliefs are safe. Presumably, we could not have easily believed the likes of $ZF + \sim\text{Con}(ZF)$!

Mathematical pluralism is radical, even in the revised form advocated. Nobody denies pluralism about (pure) geometry. Different geometries can all be realized in a single metatheory, like (some axiomatization of) set theory. Set-theoretic pluralism, by contrast, precludes any such stable background arena. Every set theory *is* a metatheory, with its own interpretation of, say, higher-order consequence.[73] So mathematical pluralism engenders a kind of *perspectivalism* about metatheoretic questions. This precludes it from being formalized. One cannot state pluralism about a potential kind of metatheory, like set theory, using a theory of that kind.[74] Any such theory would just be another metatheory and

---

[73] See Hamkins [2012, Sec. 5].

[74] One can, of course, state less radical forms of pluralism as formal theories. See, for instance, Väänänen [2014]

would take itself to be maximal.[75] On the other hand, not even the pluralist, understood as above, is so radical as to tolerate any mathematical beliefs, even if in reflective equilibrium. For instance, Nelson and Zeilberger's beliefs (Section 1.4) are false, according to the pluralist. And we have yet to consider heretic logicians.[76]

3.6 Conclusion

I have discussed the reliability challenge for mathematical realism, due to Benacerraf and Field. I have substantially clarified the dialectic and argued that the challenge is best understood as that of showing that our mathematical beliefs are *safe*, in the epistemologist's sense. Whether this challenge can be answered depends on whether we could have easily had (systematically) different beliefs. Alternatively, it depends on mathematical pluralism, the view that every consistent – or, better, arithmetically sound – foundational theory is true of its intended subject.

Could one be a pluralist about topics of philosophical controversy more generally? In particular, might one respond to the reliability challenge (understood as the challenge to show that our beliefs are safe) for realism about (counterfactual) modality, (meta)logic, or normative theory, in a similar way? I turn to this question, and its metaphilosophical ramifications, in Section 4.

4 Modality, Logic, and Normativity

I have discussed the problem of mathematical knowledge. I have partitioned it into two aspects: the challenge to explain the justification of our mathematical beliefs (the *justificatory challenge*) and the challenge to explain their reliability (the *reliability challenge*). I have considered responses to the former in terms of self-evidence, analyticity, reflective equilibrium, and scientific application. And I have substantially clarified the latter, as well as the most promising realist response to it, *mathematical pluralism*. At first approximation, mathematical pluralism says that "though mathematical objects are mind-independent, any view we had had of them would have been correct" [Field 2005, 78]. In this section, I show that analogous epistemic challenges arise across philosophy and

---

[75] It would also invite pluralism about pluralism. (This kind of metatheoretic perspectivism is similar to the ethical relativism of Rovane [2013] and the fragmentalism about time of Fine [2006]. I will sketch an extremely general version of the view, encompassing, besides evaluative areas, even logic and physics, in Sections 4.4 and 4.5.)

[76] The contemporary debate between pluralists and "objectivists" (as I call them) is reminiscent of the Frege–Hilbert Controversy, though there are also important differences between the disputes as well. See: https://plato.stanford.edu/entries/frege-hilbert/

argue that pluralism affords the most promising realist response to them as well. The upshot is a metaphysically realist, but pragmatist, metaphilosophy.

4.1 Generalizing

Recall that the *justificatory challenge* is acute for realism about an area, *F*, when the following condition is met:

(1) There is *F*-disagreement that bottoms out in conflicting intuitions.

Similarly, the *reliability challenge* is pressing for F-realism when the following holds:

(2) We could have easily had systematically different *F*-beliefs (using the method that we actually used).

Conditions (1) and (2) appear to be satisfied by areas of philosophical concern far beyond mathematics, including modal metaphysics, (meta)logic, and normative theory.[77] So, prima facie if one is a mathematical pluralist in response to the justificatory or reliability challenges, then one should be a pluralist about other areas of philosophical controversy on the same basis.

Consider modal metaphysics. This is the theory of how the world could have been. Could God have failed to exist, if God exists, in fact? Could the mereological truths – the truths governing part and whole – have been otherwise? What about the truths of identity, or, indeed, the truths of (pure) mathematics? These are modal questions. It hardly needs stating that modal metaphysics is controversial. Disagreements over every modal proposition of interest, including the aforementioned, persist.[78] Moreover, like disagreements over *AC*, modal disputes characteristically seem to bottom out in contingent intuitions. For example, while *textbook* disagreement over the claim that it is necessary that mental states are identical to brain states turn on disputes over the former's actual identity, disagreement over the *Necessity of Identity* itself, written

---

[77] The parenthetical "meta" is strictly needed before "logic" because disagreements over what follows from what, what is consistent, and so on are not about whether, for example, either every even natural number greater than 2 is equal to the sum of two prime numbers or it is not the case that every even natural number greater than 2 is equal to the sum of two prime numbers (i.e., the disjunction of *Goldbach's Conjecture* and its negation). But such disagreements *translate* into disagreements over claims like this disjunction insofar as they license different inferences. More on the connection (or lack thereof) between what follows from what and what to infer from what in Section 4.4.

[78] This is despite the fact that, like mathematics, modal metaphysics has become highly formal and quite intentionally so (Williamson [2013]). As in the mathematical case, this agreement in *practice* could give the outsider the misimpression that there is robust and reflective consensus over the truth-values of fundamentals.

$(x)(y)[x = y \rightarrow [](x = y)]$, is not thus empirical. The following quotations are representative:

> Nothing is [necessary], unless the contrary implies a contradiction. Nothing, that is distinctly conceivable, implies a contradiction. Whatever we conceive as existent, we can also conceive as non-existent. There is no being, therefore, whose non-existence implies a contradiction. Consequently there is no [God], whose existence is [necessary]. [Hume 1779, 92]

> Lewis thinks this "unrestricted composition" ... is ... necessary .... [But w]hy suppose that ... it is impossible for the world to have different principles governing the part-whole relation? [Nolan 2005, 36]

> The overriding purpose of this essay has been to supply both the motivation and the means for rejecting [the Necessity of Identity]. [Wilson 1983, 323]

> [Nominalists]s believe that numbers do not exist .... [Y]ou know perfectly well what they think .... When we work ... through the nominalist's system ... we encounter neither contradiction nor manifest absurdity .... [I]f [there is an incoherence in the view], there must be some nonmodal fact given which it is ... absurd to suppose that there might have been no numbers. But ... [w]e cannot imagine what it could ... be. [Rosen 2002, 292–4]

Metaphysicians reply to these stalemates as set theorists respond to disagreements over foundations. Philosophy is hard. There is no reason to expect that creatures evolved from great apes would have resolved recondite questions of how the world could have been. Moreover, there is a range of modal propositions over which most modal theorists do agree. Sider writes:

> [L]ogic ... is metaphysically necessary .... [L]aws of nature are not .... [I]t is metaphysically necessary that "nothing can be in two places at once," and so on. This conception falls far short of a full criterion. But a thin conception is not in itself problematic. For when a notion is taken to be fundamental, one often assumes that the facts involving the notion will outrun one's conception. [2011, 266][79]

However, those worried by the justificatory or reliability challenges will be unmoved by these assurances. Thomasson asks:

> What are we doing, when we do [modal] metaphysics?[80] A tempting answer—popular among contemporary metaphysicians—is to think of metaphysics as engaged in discovering ... fundamental facts about the world. But ... the radical and persistent disagreements that have characterized metaphysics ...

---

[79] Sider himself is a reductionist and deflationist about metaphysical possibility. See his [2011, Ch. 11] work.

[80] Thomasson is not actually focused on *modal* metaphysics here, though she is an anti-realist about it too. See her [2020] work.

lead to skepticism about whether metaphysicians are really succeeding in discovering such facts. [2016, 1]

And Stalnaker claims:

> [I]f [modal realism] were true, then it would not be possible to know any of the facts about what is ... possible .... This epistemological objection may ... parallel ... Benacerraf's dilemma [the reliability challenge] about mathematical ... knowledge. [1996, 39–405][81]

The upshot is that modal realists face justificatory and reliability challenges if mathematical realists do. We saw that mathematical pluralism affords a response. So let us explore a similar account of modality, *modal pluralism*, according to which while the truths about what is possible and necessary are independent of us, "any view we had had of them would have been correct."

4.2 Modal Pluralism

Fix a modal logic. For concreteness, we can let it be the *S5* variable domain (first-order) quantified modal logic of Priest [2008, Ch. 16] with the Necessity of Identity. Then we can achieve an initial degree of pluralism by stipulating "modal axioms" and closing under modal logical consequence. For example, we may stipulate the following (Sider [2011, 274]):

> Parthood is transitive.
> There are no (merely) past or future objects.
> Objects have temporal parts.
> Any objects have a mereological sum.

These stipulations amount to adjoining conditional axioms to the logic. Just as the *Necessity of Identity* tells us that if superman is identical to Clark Kent, then he necessarily is, we can add a principle one consequence of which is that *if it is true that if a quark is part of a proton and a proton is part of an atom, then a quark is part of an atom, then this is necessarily true*. Modal pluralism says that any collection of axioms meeting a certain condition is true of its intended subject, as in the set-theoretic case. The condition on mathematical axioms was arithmetic soundness (Section 3.5). What should the constraint be on modal axioms? The obvious one is consistency in the background modal logic. This proposal results in a hierarchy of more permissive kinds of possibility, in analogy with a hierarchy of broader kinds of set. The picture is as follows:

---

[81] Stalnaker himself seems to be under the impression that the reliability challenge for *F*-realism depends on the existence of peculiarly *F*-entities. But we saw in Section 3.1 that this is not so. In a slogan: ontology is *epistemically* irrelevant.

...

N_{-2} = technological possibility

N_{-1} = physical possibility

N_0 = metaphysical possibility

N_1 = metaphysical possibility minus the mathematical truths

N_2 = metaphysical possibility minus the mathematical and origin truths

N_3 = metaphysical possibility minus the mathematical, origin, and mereological truths

N_4 = metaphysical possibility minus the mathematical, origin, mereological,
and normative truths

...

While the Ns are totally ordered by the *more inclusive than* relation, there are incommensurate concepts of possibility, as there are incommensurate concepts of set. Let N^* = metaphysical possibility minus just the mathematical truths, and let N^{**} = metaphysical possibility minus just the origin truths. Then neither concept $< N^* >$ nor $< N^{**} >$ is more inclusive than the other.

4.3 Logical Pluralism

So far, our "modal pluralism" is really just a kind of *permissivism*. It is the view that much more is (counterfactually) possible, that is, *really could have been the case*, than we had supposed. It is analogous to the view that there is a very weak background concept of set with respect to which all others are restrictions.[82] But the reasons to deny that there is a uniquely right set of "modal axioms" are equally reasons to deny that there is a uniquely right background modal logic. The justificatory and reliability challenges arise with respect to modal logical principles too – just as they would arise with respect the axioms of a background set theory. For example, much as there are disputes over the necessity of the transitivity of parthood that bottom out in contingent intuitions, there are disputes over the *Necessity of Identity* that bottom out in these.[83]

In fact, there are disputes over *non-modal* logical principles with the same character. For example, *paracomplete* logicians reject the *Law of the Excluded*

---

[82] This is, in fact, how Field [1998] advocates formulating mathematical pluralism.

[83] Besides Wilson [1983], see Girle [2017, 7.4, 8.5, and 8.6], Gibbard [1975], and Priest [2008, Ch. 17] for modal logics with contingent identity. Did not Kripke prove that identities are necessary, appealing only to the idea of rigid designation ([Kripke 1971, 181])? If saying that names are rigid designators is to say that "Hesperus" and "Phosphorus" refer to what they actually refer to in every world, then showing that "Hesperus" and "Phosphorus" are rigid designators does not show that the terms *co-refer* at every world. It shows that "Hesperus" refers to Hesperus and that "Phosphorus" refers to Phosphorus in every world. If it means that "Hesperus" and "Phosphorus" co-refer in every world, if they do in the actual world, then Kripke assumes what he seeks to prove. See Cameron [2006].

Middle (*LEM*; i.e., they deny that $(P \vee \sim P)$ is a logical truth), while *paraconsistent* logicians allow that contradictions can – as a matter of logical possibility – be true! Williamson writes, "If we restricted [logic] to uncontroversial principles, nothing would be left" [2012]. Here are four simple alternatives to classical logic:

o *Strong Kleene, K3*: In *K*3, statements take on a third truth-value, along with truth (*T*) and falsity (*F*), commonly called *indeterminacy*, *i*. The truth-tables are such that if *P* is *i*, then so is $\sim P$, and $(P \rightarrow Q)$ is *i* if *P* or *Q* or both is. There are no logical truths in this logic, but there are valid inference rules, like *modus ponens* (from *P* and $(P \rightarrow Q)$ infer *Q*).

o *Łukasiewicz, Ł3*: The logic, Ł3, is just like *K*3, except $(P \rightarrow Q)$ is only *i* if *P* is *i* and *Q* is *F*, or if *P* is *T* and *Q* is *i*. *Ł3* has logical truths, like $(P \rightarrow P)$, as well as valid inference rules. However, *LEM* is not among them ($[P \vee \sim P]$ can fail).

o *Logic of Paradox, LP*: This is like *K*3 but with *i* regarded as a "designated value" – that is, a value that valid inferences preserve. The truth-value, *i*, is now taken to mean *both true and false*. Like *Ł3*, and unlike *K*3, this logic has logical truths. Moreover, *LEM* is among them. However, both *reductio ad absurdum* and *modus ponens* are invalid inferences.

o *First – Degree Entailment→, FDE→* : *FDE* is *both* paracomplete and paraconsistent. But when supplemented with an appropriate conditional, →, it still validates *modus ponens*.

The upshot is that modal pluralism invites logical pluralism. Indeed, if logical concepts *are* modal (as discussed in Section 2.3), then this is not even a substantive step. The claim that, say, *Q* classically follows from *P just is* the claim that it is not *classically logically possible* that *P* and $\sim Q$. Equally, it could not, as a matter of classical-logical possibility, have been that *P* and $\sim Q$.

4.4 Indefinite Extensibility and Perspectivalism

Of course, the question of where to draw the line rearises vis-à-vis logic. Is there a weakest logic with respect to which all others are mere restrictions? It might be thought that there must be. Amalgamate them! But this argument fails for two reasons. First, while it is true that for any determinate collection of logics, we can construct a weakest one by amalgamating them; the collection of *all* logics may be an indeterminate totality. Indeed, if a logic is something that we can actually *use* as an all-purpose reasoning device (rather than a mere formal object), then that totality *would* seem to be indeterminate. Beginning with any logic, we can intelligibly weaken it by considering what would have been the case had a validity in the logic failed. However, the "trivial logic," according to which everything is consistent with everything, and nothing follows from anything, is unintelligible (as an all-purpose reasoning device). The concept of a weakest logic would, thus, appear to be *indefinitely extensible* in something like the sense of Dummett (Clarke-Doane [2019a, Section.8]). "[I]f we can form a definite conception of a totality all of whose members fall under the concept [of logical consistency], we can, by reference to that totality, characterize a larger totality all of whose members fall under it" [Dummett 1993, 441].[84]

The second problem with the argument is that it forgets that the strength of a logic is perspectival, just like the breadth of a concept of set. Let us return to the set-theoretic case. Martin complains, "[t]he models postulated by [mathematical pluralist] determine a canonical maximal set-theoretic structure, the amalgamation. If one takes those models seriously, then one should regard this canonical structure as the true universe of sets" [2001, 14]. The problem with this argument, alluded to in Section 3.5, is that it assumes that we can stably compare competing set concepts. In order to make the comparison, we need a metatheory, which will itself use a concept of set! So, for the pluralist, the only lesson is that "*within any fixed set-theoretic background*" ([Hamkins 2012, 427, italics in original]), there is a broadest concept of set. It is not that there is a broadest concept of set *period*. In a similar way, what follows from what, what is consistent, and what is a logical truth in a logic *depend on the logic one uses to check* (Shapiro [2014, Ch. 7]). Logic, L, may fail to be the weakest logic, relative to some logics, even if *relative to some other logic* it is.

An attraction of modal and logical pluralism is that it is not even clear that there is a sensible alternative – unless this is just a thesis about natural language semantics.[85] In the mathematical case, there is at least a prima facie disagreement over ontology. Do there only exist *ZFC* sets? Or do there also exist sets (or set-like things) some of which lack well-orders? But there need be no *onto-*

---

[84] Dummett is actually talking about sets of mathematical objects here. (Note that unlike the case of mathematical objects, one cannot explain the indefinite extensibility of logical possibility – or consistency, understood as a modal idea – modally. That would be too close to circular. So, if we are looking for a uniform account of indefinite extensibility and agree that absolute possibility is indefinitely extensible, then we should reject what are perhaps the most salient accounts of the phenomenon in terms of modal language like Dummet's "*can* ... characterize.")

[85] That is, pluralism is compatible with the semantic hypothesis that we all use the words "follows," "entails," and so on determinately to mean one thing (just as geometric pluralism is compatible with the hypothesis that we use "point," "line," etc. to mean, e.g., Euclidean point, line, etc., and the relativity of simultaneity is compatible with the hypothesis that we use 'simultaneous' to mean simultaneous-relative-to-reference-frame-R).

logical dispute in the modal or logical cases.[86] One could try to argue that an ideological quarrel (in the sense of Quine [1951b]) remains. But who denies that we can *stipulatively introduce* consequence relations? Maybe most of these are not "real" consequence relations. But what could this mean, if not just something about how we happen to use the word "consequence"? We speak a language, and, for all that has been said, we may mean classical consequence by "consequence." In that case, it may be mind- and language-independently true that *LP* consequence is not real consequence and that classical consequence is. (Trivially, "consequence" refers to consequence, so anything that does not refer to this is not real consequence.) But it does not follow that *LP* consequence does not "exist," much less that we ought not use it to reason! Indeed, it is hard to think of *anything* of interest that follows from a logic's failing to be real in this sense. If the claim that, say, classical consequence is real *were* the claim that "the ... relation so defined agrees with the pre-theoretic notion of implication between statements" [Zach 2018, 2080], then Riemannian lines may not be real![87]

The picture that results from combining modal, logical, and mathematical pluralism is dizzying. There is a pluriverse of sets, whose nature is relative to a metatheory. That pluriverse is also an indefinitely extensible – not just perspectival – totality, since the notions of set and consistency are.[88] When physical science is taken into account, things get wilder. A physical theory is *defined* to be the closure of some principles under a logic. Since logics abound, any theory, such as the Standard Model, really corresponds to a plethora of theories, for each of the logics under which one could close its principles. There is the closure of the Standard Model's Lagrangian (density) under classical logic, the Logic of Paradox, First Degree Entailment, and everything in between. Meanwhile, the various kinds of *possibility* engender different state spaces

---

[86] Unless one is a Lewisian, the ontology in question, like sentences, is not peculiar to modality or logic. It turns on generic questions about the existence of universals.

[87] It might be thought that one could appeal to the concept of naturalness (or "metaphysical privilege") in the sense of Sider [2011] in order to give content to the claim that some consequence relation is real. But either the claim that some consequence relation is natural is itself a normative claim (with implications for how we *ought* to reason) or not. If it is, then the problem of pluralism rearises vis-à-vis naturalness (Section 4.5). The consequence relation will be, for example, natural$_{classical}$ but not natural$_{LP}$. What could the claim that naturalness$_{classical}$ is real naturalness amount to if not just something about how we use "naturalness"? Alternatively, the claim that some consequence relation is natural is not normative. But then it is neither here nor there from the standpoint of which to use, for familiar is/ought reasons. See Clarke-Doane [2020b, 6.6].

[88] That the notion of set (as opposed to consistency) is indefinitely extensible is commonly accepted even by "monists" about set theory (thanks to *Russell's Paradox*, the *Burali-Forti Paradox*, etc.). This normally connotes the "height" of the set-theoretic hierarchy. See Shapiro and Wright [2006]. But it can also be argued that the notion of set is indefinitely extensible by "width" using forcing. See n. 55.

(even fixing on a modal logic). So the multitude of Standard Models include different sentences, countenance different states of, for example, a weak isospin system, and, assuming mathematical pluralism, are governed by different set-theoretic laws! The Standard Model becomes a "cloud" of triples, given by a choice of set theory, modality, and (meta)logic. Insofar as mathematical, modal, and logical reality is multifarious and indeterminate, the physical universe must be too.

4.5 Realist Carnapianism

If one rides the pluralist train to the last station, then one arrives at realism with a pragmatist spin. The world – including its mathematical, modal, or logical aspects – is out there, independent of us. We do not make up the pluriverse of sets, possibilities, or consequence relations, despite their perspectival and indefinite character.[89] Realism is true of all three subjects. Nevertheless, for practical purposes, it is *as if* realism about them were false and conventionalism were true.

Consider, again, the Parallel Postulate. If mathematical realism is true, then how things are with the (pure) geometrical points and lines is independent of minds and languages. We do not make up Euclidean, hyperbolic, elliptic, or variably curved spaces, any more than we make up the natural numbers. And, yet, a dispute over the Parallel Postulate (understood as pure mathematics) would be patently misguided. All we would learn by resolving it is something about ourselves. We would just learn which geometrical structures we were talking about, rather than learning which ones there were. This is why, in practice, we simply stipulate that we will use "point" and "line" to mean, say, Euclidean point and line. It is *as if* our conventions *made* the axioms true. It is as if the axioms of (pure) geometry were *metaphysically* analytic (Section 1.3).

If pluralism is true, then this is the much more general situation. The question of whether *AC* is true is like that of whether the Parallel Postulate is. There is nothing of *metaphysical* significance at stake. Perhaps conceptual surgery reveals that *AC* is "built into" the concept of set that we happen to have inherited. It does not follow that there are no sets – or set-like things – that lack well-orders![90] The same is true of paradigmatic modal and logical questions. Could mental states have failed to be identical to brain states? They could

---

[89] In the logical case, this means that we neither make up which logic is correct, nor what follows from what in the correct logic. This distinction was drawn in the Introduction.

[90] Thus, if pluralism is true, then the "[m]any set theorists, including Gödel" who "believe that conceptual analysis will eventually lead to an idea of a set so clear and distinct that the answer to the continuum question will become apparent" [Huber-Dyson 1991, 9] are misguided *even if* their belief is true.

have as a matter of logical possibility, and they could not have as a matter of possibility that builds in the psycho-physical laws! There is no fact left to dispute except what we happen to mean by "could have."[91]

It might be thought that at least questions of applied logic resist deflation, as questions of applied geometry do. Indeed, Putnam claims that "It makes as much sense to speak of 'physical logic' as of 'physical geometry'. We live in a world with a non-classical [quantum] logic" [1968, 226]. But Putnam never tells us what this means. In the geometric case, it means that initially parallel geodesics (e.g., light rays) do not stay the same distance apart. But what is the physical content of the key claim   of   quantum   logic   that   the   *Distributive   Law*, $P \& (Q \vee R) \longleftrightarrow [(P \& Q) \vee (P \& R)]$, is invalid? Let S_i and T_i be eigenstates of position and momentum, respectively. Then, according to quantum logic: $(S_1 \vee S_2 \vee \ldots S_j) = (T_1 \vee T_2 \vee \ldots T_k) = $ span of spaces $= \top$, while $(S_l \& T_m) = $ intersection of spaces $= \bot$, for all l and m. So, indeed, $S_i \& (T_1 \vee T_2 \vee \ldots T_k) = S_i \neq (S_i \& T_1) \vee (S_i \& T_2) \vee \ldots = \bot \vee \bot \ldots = \bot$. Still, classical consequence (respecting the distributive law) "exists" if quantum consequence does! So there is no metaphysical question at stake.[92] The only sensible view is that quantum consequence is more useful for modeling Yes/No questions in quantum mechanics. This is like the (standard) view that while Euclidean, hyperbolic, and other spaces exist, if any do, *as pure mathematical structures*, a certain (pseudo-)Riemannian space is most useful for modeling space-time. *Unlike* the geometric case, however, there is no metaphysical remainder in the logical case – no physical analog to lines.

The upshot is that the nonverbal questions in set theory, modal metaphysics, and logic are *normative*. *Ought* we use the iterative hierarchical concept of set, the metaphysical concept of possibility, or the classical concept of consequence for the purpose at hand? Actually, not even this gets to the bottom of things. Normative questions, factually construed, admit of the same pluralist deflation. We ought$_{quantum}$ reason according to quantum logic, ought$_{classical}$ reason according to classical logic, and so on. Similarly, there are norms, ought$_1$ and ought$_2$, according to which we ought$_1$ adopt $V = L$, but ought$_2$ not. Just as it makes little

---

[91] This means that influential modal arguments for conclusions about the actual world must be too quick. Consider the standard argument for dualism (Chalmers [1996]). Even if it is conceivable that mental states are not physical states, *and even if this shows that they really could have been distinct*, it does not show that they could have been distinct *in a sense that satisfies the Necessity of Identity*. Given that metaphysical possibility is just one among countless kinds, the "worlds" in which mental states fail to be brain states may lie outside the class of worlds for which the Necessity of Identity holds. Whether one thinks so will turn on whether one thinks that mental states are actually physical states. See Clarke-Doane [2020c].

[92] It is hard to see how changing the logic would help with the Measurement Problem, even if it were metaphysical. The problem could just be rephrased. Why does measuring an *indeterminate determinable* result in a *determinate*?

sense to say that, for example, quantum consequence "exists" to the exclusion of classical consequence, it makes little sense to say that quantum ought exists to the exclusion of classical ought. (And since the justificatory and reliability challenges arise equally for normative realism [Enoch (2009), Huemer (2005, 99)], there would be an argument for normative pluralism like the argument for mathematical pluralism even if normative "monism" made sense.) The only *factual* question in the vicinity is what language we happen to speak. But *what to infer from what* – and, generally, *what to do* – is not be resolved by natural language semantics (Clarke-Doane [2020b, Ch. 6])!

When the dust settles, pluralism mimics Carnap [1950], despite its antithetical metaphysics. The question of which mathematical, modal, or logical axioms are true *is* misconceived, as Carnap alleged. And the pressing questions in the vicinity are nonfactual practical questions of what to do. What concept of set to use? What concept of possibility and consequence to employ? Indeed, what concept of "ought" to follow – where this is *not* the (circular) question of what concept of ought we *ought* to follow (Clarke-Doane [2021]). Theoretical questions dissolve into practical ones, questions of expedience. As Carnap writes, "the conflict between the divergent points of view ... disappears ... [B]efore us lies the boundless ocean of unlimited possibilities" [1937/2001, XV].

Conclusions

I have discussed the problem of mathematical knowledge. I have argued that the justificatory challenge, the challenge to explain the defeasible justification of our mathematical beliefs, arises insofar as disagreement over axioms bottoms out in disagreement over intuitions. And I have argued that the reliability challenge, the challenge to explain their reliability, arises to the extent that we could have easily had different beliefs. I showed that mathematical facts are not, in general, empirically accessible, contra Quine, and that they cannot be dispensed with, contra Field. However, I argued that they might be so plentiful that our knowledge of them is intelligible. I concluded by sketching a complementary "pluralism" about modality, logic, and normative theory. I highlighted its surprising metaphysical and methodological ramifications. Metaphysically, pluralism engenders a kind of perspectivalism and indeterminacy. Methodologically, pluralism vindicates Carnap's pragmatism, transposed to the key of realism.

References

Arrigoni, Tatiana. [2011] "V=L and Intuitive Plausibility in Set Theory. A Case Study." *Bulletin of Symbolic Logic*. Vol. 17. 337–359.

and Sy Friedman. [2012] "Foundational Implications of the Inner Model Hypothesis." *Annals of Pure and Applied Logic*. Vol. 163. 1360–1366.

Arntzenius, Frank, and Cian Dorr. [2012] "Calculus as Geometry." In Arntzenius, Frank (ed.), *Space, Time and Stuff*. Oxford: Oxford University Press. 213–68.

Azcel, Peter. [1988] Non-Well-Founded Sets. SLI Lecture Notes. Vol. 14. Stanford, CA: Stanford University, Center for the Study of Language and Information.

Balaguer, Mark. [1995] "A Platonist Epistemology." *Synthese*. Vol. 103. 303–325.

[1996] "Toward a Nominalization of Quantum Mechanics." *Mind*. Vol. 105, No. 418. 209–226.

[2001] "A Theory of Mathematical Correctness and Mathematical Truth." *Pacific Philosophical Quarterly*. Vol. 82. 87–114.

Baras, Dan and Justin Clarke-Doane. [2021] "Modal Security." *Philosophy and Phenomenological Research*. Vol. 102, No. 1. 162–183.

Barton, Neil. [2016] "Multiversism and Concepts of Set: How Much Relativism is Acceptable?" In Boccuni, Francesca and Andrea Sereni (eds.), *Objectivity, Realism, and Proof*. Switzerland: Springer, 189–209.

Bell, John, and Geoffrey Hellman. [2006] "Pluralism and the Foundations of Mathematics." In Waters, Kenneth, Helen Longino, and Stephen Kellert (eds.), *Scientific Pluralism* (Minnesota Studiesin Philosophy of Science, Volume 19). Minneapolis: University of Minnesota Press, 64–79.

Benacerraf, Paul. [1965] "What Numbers Could Not Be." *Philosophical Review*. Vol. 74. 47–73.

[1973] "Mathematical Truth." *Journal of Philosophy*. Vol. 70. 661–679.

Bengson, John. [2015] "Grasping the Third Realm." Gendler, Tamar Szabó and John Hawthorne (eds.), *Oxford Studies in Epistemology*, Vol. 5. Oxford: Oxford University Press, 1–34.

Berto, Francesco. [2009] There's Something about Gödel: The Complete Guide to the Incompleteness Theorem. Oxford: Wiley-Blackwell.

Boolos, George. [1971] "The Iterative Conception of Set." *Journal of Philosophy*. Vol. 68. 215–231.

[1999] "Must We Believe in Set Theory?" In Jeffery, Richard (ed.), *Logic, Logic, and Logic*. Cambridge, MA: Harvard University Press, 120–133.

Boghossian, Paul. [2003] "Epistemic Analyticity: A Defense." *Grazer Philosophische Studien*. Vol. 66. 15–35.

Bourbaki, Nicolas. [1970] Elements of Mathematics: Theory of Sets. Heidelberg: Springer-Verlag.

Butterworth, Brian. [1999] What Counts? How Every Brain is Hardwired for Math. New York: The Free Press.

Bueno, Otávio. [2020] "Nominalism in the Philosophy of Mathematics." *The Stanford Encyclopedia of Philosophy* (Fall 2020 ed.), Zalta, Edward N. (ed.), https://plato.stanford.edu/archives/fall2020/entries/nominalism-mathematics/

Cameron, Ross. [2006] "Comment on 'Kripke's (Alleged) Argument for the Necessity of Identity Statements'." *Wo's Weblog*. www.umsu.de/wo/archive/2006/08/09/Kripke_s__Alleged__Argument_for_the_Necessity_of_Identity_Statements

Carnap, Rudolf. [1950] "Empiricism, Semantics, and Ontology." *Revue Internationale de Philosophie*. Vol. 4. 20–40. Reprinted in the Supplement to Meaning and Necessity: A Study in Semantics and Modal Logic,enlarged edition (University of Chicago Press, 1956). https://tu-dresden.de/gsw/phil/iphil/theor/ressourcen/dateien/braeuer/lehre/meta meta/Carnap–EmpiricismSemanticsOntology.pdf?lang=en

[2010/1937] The Logical Syntax of Language. Oxford: Routledge.

Carey, Susan. [2009] The Origin of Concepts. Oxford: Oxford University Press.

Chalmers, David. [1996] The Conscious Mind. New York: Oxford University Press.

Chen, Eddy Keming. [2019] "The Intrinsic Structure of Quantum Mechanics." *Essays on the Metaphysics of Quantum Mechanics*. PhD dissertation. New Brunswick: Rutgers University.

Cheyne, C. [1998]. "Existence Claims and Causality." *Australasian Journal of Philosophy*. Vol. 76. 34–47.

Chudnoff, Elijah. [2013] Intuition. Oxford: Oxford University Press.

Clarke-Doane, Justin. [2012] "Morality and Mathematics: The Evolutionary Challenge," *Ethics*. Vol. 122. 313–340.

[2015] "Justification and Explanation in Mathematics and Morality." Shafer-Landau, Russ (ed.), *Oxford Studies in Metaethics*, Vol. 10. New York: Oxford University Press.

[2016] "What is the Benacerraf Problem?" In Pataut, Fabrice (ed.), *New Perspectives on the Philosophy of Paul Benacerraf: Truth, Objects, Infinity*. Dordrecht: Springer, 17–43.

[2019a] "Modal Objectivity." *Noûs*. Vol. 53. 266–295.

[2019b] "Metaphysical and Absolute Possibility." *Synthese* (Suppl 8). Vol. 198. 1861–1872.

[2020a] "Set-theoretic Pluralism and the Benacerraf Problem." *Philosophical Studies*. Vol. 177. 2013–2030.

[2020b] Morality and Mathematics. Oxford: Oxford University Press.

[2020c] "Undermining Belief in Consciousness," for an author-meets-critics symposium on David Chalmers' "The Meta-Problem of Consciousness," with replies from Chalmers, *Journal of Consciousness Studies*. Vol. 26. 34–47.

[2021] "From Non-Usability to Non-Factualism," for an author-meets-critics symposium on Holly Smith's *Making Morality Work*, with replies from Smith, *Analysis*. Vol. 81. 747–758.

Cohen, Paul. [1966] Set Theory and the Continuum Hypothesis. New York: W. A. Benjamin.

[1971] "Comments on the Foundations of Set Theory." In Scott, Dana (ed.), *Axiomatic Set Theory* (Proceedings of Symposia of Pure Mathematics, Vol. XIII, Part I). Providence: American Mathematical Society.

Colyvan, Mark. [2007] "Mathematical Recreation versus Mathematical Knowledge." In Leng, M. A. Paseau, and M. Potter (eds.), *Mathematical Knowledge*. Oxford: Oxford University Press, 109–122.

De Cruz, Helen. [2006] "Why are Some Numerical Concepts More Successful than Others? An Evolutionary Perspective on the History of Number Concepts." *Evolution and Human Behavior*. Vol. 27. 306–323.

Dehaene, Stanislas. [1997] The Number Sense: How the Mind Creates Mathematics. Oxford: Oxford University Press.

De Toffoli, Silvia. [2021] "Groundwork for a Fallibilist Account of Mathematics." *Philosophical Quarterly*. 7(4). 823–844.

Devlin, Keith. [1977] The Axiom of Constructability: A Guide for the Mathematician (Lecture Notes on Mathematics: 617). New York: Springer-Verlag.

[1981] "Infinite Trees and the Axiom of Constructibility." *Bulletin of the London Mathematical Society*. Vol. 13. 193–206.

Enoch, David. [2009] "The Epistemological Challenge to Metanormative Realism: How Best to Understand It, and How to Cope with It." *Philosophical Studies*. Vol. 148. 413–438.

Eskew, Monroe. [2019] "Re: Why Not Adopt the Constructability Axiom?" May 20. https://mathoverflow.net/questions/331956/why-not-adopt-the-constructibility-axiom-v-l

Feferman, Solomon. [1992] "Why a Little Bit Goes a Long Way: Logical Foundations of Scientifically Applicable Mathematics." *Proceedings of the Philosophy of Science Association*. Vol. 2. 442–455.

Ferrier, Edward. [2019] "Against the Iterative Conception of Set." *Philosophical Studies*. Vol. 176. 2681–2703.

Field, Hartry. [1980] Science without Numbers. Princeton: Princeton University Press.

[1989] Realism, Mathematics, and Modality. Oxford: Blackwell.

[1991] "Modality and Metalogic." *Philosophical Studies*. Vol. 62. 1–22.

[1996] "The A Prioricity of Logic." *Proceedings of the Aristotelian Society*. Vol. 96. 359–379.

[1998] "Which Mathematical Undecidables Have Determinate Truth-Values?" In Dales, H. Garth, and Gianluigi Oliveri (eds.), *Truth in Mathematics*. Oxford: Oxford University Press, 291–310.

[2005] "Recent Debates about the A Priori." In Gendler, Tamar and John Hawthorne (eds.), *Oxford Studies in Epistemology*, Vol. 1. Oxford: Clarendon Press, 69–88.

Dummett, Michael. [1993] The Seas of Language. Oxford: Oxford University Press.

Faraci, David. [2019] "Groundwork for an Explanationist Account of Epistemic Coincidence." *Philosophers' Imprint*. Vol. 19. 1–26.

Fine, Kit. [2006] "The Reality of Tense." *Synthese*. Vol. 150. 399–414.

Fontanella, Laura. [2019] "How to Choose New Axioms for Set Theory?" In S. Centrone et al., (eds.), *Reflections on the Foundations of Mathematics*, Synthese Library 407, Springer Verlag, 27–42. https://doi.org/10.1007/978-3-030-15655-8_2

Forster, Thomas. [Forthcoming] The Axioms of Set Theory. Cambridge: Cambridge University Press. Available online (in preparation): www.dpmms.cam.ac.uk/~tf/axiomsofsettheory.pdf

Fraenkel, Abraham, Yehoshua Bar-Hillel, and Azriel Levy. [1973] Foundations of Set Theory (Studies in Logic and the Foundations of Mathematics, Volume 67). New York: Elsevier Science.

Frege, Gottlob. [1980/1884] The Foundations of Arithmetic: A Logico-Mathematical Inquiry into the Concept of Number (2nd rev. ed.). Austin, J. L. (trans.). Evanston: Northwestern University Press.

Friedman, Harvey. [1973] "The Consistency of Classical Set Theory Relative to a Set Theory with Intuitionistic Logic." *Journal of Symbolic Logic*. Vol. 38. 315–319.

[2000] "Re: FOM: Does Mathematics Need New Axioms?" Post on the Foundations of Mathematics (FOM) Listserv. May 25, 2000. www .personal.psu.edu/t20/fom/postings/0005/msg00064.html

[2002] Philosophical Problems in Logic (Lecture Notes). Princeton University. https://cpb-us-w2.wpmucdn.com/u.osu.edu/dist/1/1952/files/2014/01/Princeton532-1pa84c4.pdf

Gaifman, Haim. [2012] "On Ontology and Realism in Mathematics." *The Review of Symbolic Logic*. Vol. 5. 480–512.

Gödel, Kurt. [1990/1938] "The Consistency of the Axiom of Choice and the Generalized Continuum Hypothesis." In Feferman, Solomon (ed.), *Godel's Collected Works*, Vol. II. New York: Oxford University Press, 33–102.

[1947] "What is Cantor's Continuum Problem?" *American Mathematical Monthly*. Vol. 54. 515–525.

[1990/1947] "Russell's Mathematical Logic." In Feferman, Solomon (ed.), *Gödel's Collected Works*, Vol. II. New York: Oxford University Press, 119–144.

Goldman, Alvin. [1967] "A Causal Theory of Knowing." *The Journal of Philosophy*. Vol. 64, No. 12. 357–372.

Goodman, Nelson. [1955] Fact, Fiction, and Forecast. Cambridge, MA: Harvard University Press.

Greene, Joshua. [2013] Moral Tribes: Emotion, Reason, and the Gap between Us and Them. New York: Penguin.

Gibbard, Alan. [1975] "Contingent Identity." *Journal of Philosophical Logic*. Vol. 4. 187–221.

Girle, Rod. [2017] Modal Logics and Philosophy (2nd ed.). Chicago: McGill-Queens University Press.

Guedj, Denis. [1985] "Nicholas Bourbaki, Collective Mathematician: An Interview with Claude Chevalley." *The Mathematical Intelligencer*. Vol. 7. 18–22. Translated by Jeremy Gray. www.ocf.berkeley.edu/~lekheng/interviews/ClaudeChevalley.pdf

Hager, Amit. [2014] Discrete or Continuous?: The Quest for Fundamental Length in Modern Physics. Cambridge: Cambridge University Press.

Hamkins, Joel David. [2012] "The Set-Theoretic Multiverse." *Review of Symbolic Logic*. Vol 5. 416–449.

[2014] "Re: Is there any Research on Set Theory without Extensionality Axiom?" Post on MathOverflow. May 27, 2014. https://mathoverflow.net/questions/168287/is-there-any-research-on-set-theory-without-extensionality-axiom

Harris, Michael. [2015] Mathematics without Apologies: Portrait of a Problematic Vocation. Princeton: Princeton University Press.

Hart, W. D. [1996] "Introduction." In Hart, W. D. (ed.), *The Philosophy of Mathematics*. Oxford: Oxford University Press.

Hilbert, David. [1983/1936] "On the Infinite." In Benacerraf, Paul, and Hilary Putnam (eds.), *Philosophy of Mathematics: Selected Readings* (2nd ed.). Cambridge: Cambridge University Press, 183–202.

Huber-Dyson, Verena. [1991] Gödel's Theorems: A Workbook on Formalization. Vieweg: Teubner Verlag.

Huemer, Michael. [2005] Ethical Intuitionism. New York: Palgrave Macmillan.

Hume, David. [1779] "Dialogues Concerning Natural Religion." https://quod .lib.umich.edu/cgi/t/text/pageviewer-idx?cc=ecco;c=ecco;idno=004895521 .0001.000;node=004895521.0001.000:2.10;seq=94;page=root;view=text

Jensen, Ronald. [1995] "Inner Models and Large Cardinals." *Bulletin of Symbolic Logic*. Vol. 1. 393–407.

Joyce, Richard. [2008] "Precis of the Evolution of Morality." *Philosophy and Phenomenological Research*. Vol. 77. 213–218.

Joyce, Richard. [2016] "Evolution, Truth-Tracking, and Moral Skepticism," in *Essays in Moral Skepticism*. Oxford: Oxford University Press.

Kennedy, Juliette and Mark Van Atten. [2003] "On the Philosophical Development of Kurt Gödel." *The Bulletin of Symbolic Logic*. Vol. 9. 425–476.

"Kurt Gödel." *The Stanford Encyclopedia of Philosophy* (Winter 2020 ed.), Zalta, Edward N. (ed.), https://plato.stanford.edu/archives/win2020/ entries/goedel/

Kilmister, Clive W. [1980] "Zeno, Aristotle, Weyl and Shuard: Two-and-a-Half Millennia of Worries over Number." *Mathematical Gazette*. Vol. 64. 149–158.

Koellner, Peter. [2014] "On the Question of Absolute Undecidability." *Philosophia Mathematica*. Vol. 14. 153–188.

"Large Cardinals and Determinacy." *The Stanford Encyclopedia of Philosophy* (Spring 2014 ed.), Zalta, Edward N. (ed.), https://plato .stanford.edu/archives/spr2014/entries/large-cardinals-determinacy/

Kreisel, George. [1967a] "Observations on Popular Discussions of the Foundations of Mathematics." In Scott, Dana. (ed.), *Axiomatic Set Theory* (Proceedings of Symposia in Pure Mathematics, V. XIII, Part I). Providence: American Mathematical Association, 189–198.

[1967b] "Informal Rigor and Completeness Proofs". In Imre Lakatos (ed.), *Problems in the Philosophy of Mathematics*. Amsterdam: North-Holland, 138–171.

McCarthy, William & Justin Clarke-Doane. [Forthcoming] "Modal Pluralism and Higher-Order Logic." *Philosophical Perspectives*.

Katz, Jerrold. [2002] "Mathematics and Metaphilosophy." *Journal of Philosophy*. Vol. 99. 362–390.

Kripke, Saul. [1971] "Identity and Necessity." In Munitz, Milton K. (ed.), *Identity and Individuation*. New York: New York University Press, 161–191.

Leng, Mary. [2007] "What's There to Know?" In Leng, Mary, Alexander Paseau, and Michael Potter (eds.), *Mathematical Knowledge*. Oxford: Oxford University Press, 84–108.

[2009] "'Algebraic' Approaches to Mathematics." In Otavio Bueno and Øystein Linnebo (eds.), *New Waves in the Philosophy of Mathematics*. New York: Palgrave Macmillan, 117–134.

[2010] Mathematics and Reality. Oxford: Oxford University Press.

Lewis, David. [1983] "New Work for a Theory of Universals." *Australasian Journal of Philosophy*. Vol. 61. 343–377.

[1986] On the Plurality of Worlds. Oxford: Blackwell.

Liggins, David. [2010] "Epistemological Objections to Platonism." *Philosophy Compass*. Vol. 5. 67–77.

Linnebo, Øystein. [2006] "Epistemological Challenges to Mathematical Platonism." *Philosophical Studies*. Vol. 129. 545–574.

Maddy, Penelope. [1988a] "Believing the Axioms: I." *Journal of Symbolic Logic*. Vol. 53. 481–511.

[1988b] "Believing the Axioms: II." *Journal of Symbolic Logic*. Vol. 53. 736–764.

[1997] Naturalism in Mathematics. Oxford: Clarendon Press.

Magidor, Menachem. [2012] "Some Set Theories are More Equal." Unpublished notes, http://logic.harvard.edu/EFIMagidor.pdf

Marcus, Russell. [2017] Autonomy Platonism and the Indispensability Argument. New York: Lexington Books.

Marshall, Oliver. [2017] "The Psychology and Philosophy of Natural Numbers." *Philosophia Mathematica*. Vol. 26. 40–58.

Martin, D. A. [1976] "Hilbert's First Problem: The Continuum Hypothesis." In Browder, Felix (ed.), *Mathematical Developments Arising from Hilbert Problems* (Proceedings of Symposia in Pure Mathematics. Vol. 28). Providence: American Mathematical Society, 81–93.

[1998] "Mathematical Evidence." In Dales, H. G. and G. Oliveri (eds.), *Truth in Mathematics*. Oxford: Clarendon, 215–231.

Mayberry, John. [2000] The Foundations of Mathematics in the Theory of Sets. Cambridge: Cambridge University Press.

Merritt, David. [2020] A Philosophical Approach to MOND: Assessing the Milgromian Research Program in Cosmology. Cambridge: Cambridge University Press.

Milgrom, Mordehai. [2002] "Does Dark Matter Really Exist?" *Scientific American.* Vol. 287, No. 2 (August 2002). 42–46, 50–52.

Mill, John Stuart. [2009/1882] A System of Logic, Ratiocinative and Inductive (8th ed.). New York: Harper & Brothers. www.guten-berg.org/files/27942/27942-pdf.pdf

Mogensen, Andreas. [2016] "Disagreement in Moral Intuition as Defeaters." *Philosophical Quarterly.* Vol. 67, No. 267. 282–302.

Montero, Barbara. [2019] "Benacerraf's Nonproblem." *The CUNY Logic and Metaphysics Workshop.* https://logic.commons.gc.cuny.edu/2019/10/12/benacerrafs-non-problem-barbara-gail-montero/

Nelson, Edward. [1986] Predicative Arithmetic (Mathematical Notes. No. 32). Princeton: Princeton University Press.

   [2013] "Re: Illustrating Edward Nelson's Worldview with Nonstandard Models of Arithmetic." Post on Math Overflow. October 31. https://mathoverflow.net/questions/142669/illustrating-edward-nelsons-worldview-with-nonstandard-models-of-arithmetic

Newton, Isaac. [2007] "Original Letter from Isaac Newton to Richard Bentley." *The Newton Project.* www.newtonproject.ox.ac.uk/view/texts/normalized/THEM00258

Nolan, Daniel. [2005] The Philosophy of David Lewis. New York: Routledge.

Opfer, John, Richard Samuels, Stewart Shapiro and Eric Snyder. [2021] "Unwarranted Philosophical Assumptions in Research on ANS." *Behavioral and Brain Sciences.* Vol. 44. https://doi.org/10.1017/S0140525X21001060

Pantsar, Markus. [2014] "An Empirically Feasible Approach to the Epistemology of Arithmetic." *Synthese.* Vol. 191. 4201–4229.

Potter, Michael. [2004] Set Theory and Its Philosophy: A Critical Introduction. Oxford: Oxford University Press.

Pryor, James. [2000] "The Skeptic and the Dogmatist." *Noûs.* Vol. 34. 517–549.

Putnam (eds.), Philosophy of Mathematics: Selected Readings (2nd ed.). Cambridge: Cambridge University Press.

Priest, Graham. [2008] An Introduction to Non-Classical Logic: From If to Is (2nd ed.). New York: Cambridge University Press.

Pudlák, Pavel. [2013] Logical Foundations of Mathematics and Computational Complexity Theory: A Gentle Introduction. New York: Springer.

Putnam, Hilary. [1965] "Craig's Theorem." *Journal of Philosophy.* Vol. 62. 251–260.

[1967] "Mathematics without Foundations." *Journal of Philosophy*. Vol. 64. 5–22.

[1968] "Is Logic Empirical?" In Cohen, Robert S. and Marx W. Wartofsky (eds.), *Boston Studies in the Philosophy of Science*, Vol. 5. Dordrecht: D. Reidel, 216–241.

[1980] "Models and Reality." *Journal of Symbolic Logic*. Vol. 45. 464–482.

[2012] "Indispensability Arguments in the Philosophy of Mathematics." In De Caro, Mario, and David Macarthur. (eds.), *Philosophy in the Age of Science: Physics, Mathematics, Skepticism*. Cambridge, MA: Harvard University Press, 181–201.

Quine, W. V. O. [1937] "New Foundations for Mathematical Logic." *American Mathematical Monthly*. Vol. 44. 70–80.

[1948] "On What There Is." *Review of Metaphysics*. Vol. 2. 21–38.

[1951a] "Two Dogmas of Empiricism." *Philosophical Review*. Vol. 60. 20–43.

[1951b] "Ontology and Ideology." *Philosophical Studies*. Vol. 2. 11–15.

[1969] Set Theory and Its Logic (rev. ed.). Cambridge, MA: Harvard University Press.

[1986] "Reply to Charles Parsons." In Hahn, Lewis Edwin and Paul Arthur Schilp (eds.), *The Philosophy of W. V. Quine* (Library of Living Philosophers, Volume 18). La Salle: Open Court.

Rabin, Gabriel. [2007] "Full-Blooded Reference." *Philosophica Mathematica*. Vol. 15. 357–365.

Rawls, John. [1971] A Theory of Justice. Cambridge, MA: Harvard University Press.

[1974] "The Independence of Moral Theory." *Proceedings and Addresses of the American Philosophical Association*. Vol. 47. 5–22.

Relaford-Doyle, Josephine and Rafael Núñez. [2018] "Beyond Peano: Looking into the Unnaturalness of the Natural Numbers." In Bangu, Sorin (ed.), *Naturalizing Logico-Mathematical Knowledge Approaches from Philosophy, Psychology and Cognitive Science*. New York: Routledge, 234–251.

Restall, Greg. [2003] "Just What is Full-Blooded Platonism?" *Philosophia Mathematica*. Vol. 11. 82–91.

Rieger, Adam. [2011] "Paradox, ZF, and the Axiom of Foundation" In DeVidi, David, Hallet, Michael, and Peter Clark (eds.), *Logic, Mathematics, Philosophy, Vintage Enthusiasms: Essays in Honour of John L. Bell* (The Western Ontario Series in Philosophy of Science). New York: Springer, 171–187.

Rosen, Gideon. [2002] "A Study of Modal Deviance." In Tamar Szabo Gendler and John Hawthorne (eds.), *Conceivability and Possibility*. Oxford: Clarendon, 283–309.

Rovane, Carol. [2013] The Metaphysics and Ethics of Relativism. Cambridge, MA: Harvard University Press.

Rovelli, Carlo. [2007] Quantum Gravity. Cambridge: Cambridge University Press.

Russell, Bertrand. [1907] "The Regressive Method for Discovering the Premises of Mathematics." In Lackey, Douglas (ed.), [1973] *Essays in Analysis*, by Bertrand Russell. London: Allen and Unwin, 272–283.

[1918] "The Philosophy of Logical Atomism." *The Monist.*Vol. XXVIII.

Scott, Dana. [1961] "More on the Axiom of Extensionality." In Bar-Hillel, Y., E. I. J. Poznanski, M. O. Rabin, and A. Robinson (eds.), *Essays on the Foundations of Mathematics*, Dedicated to A. A. Fraenkel on his Seventieth Anniversary. Jerusalem: Magnes Press, 495–527.

Schechter, Joshua. [2010] "The Reliability Challenge and the Epistemology of Logic." *Philosophical Perspectives*. Vol. 24. 437–464.

Shapiro, Stewart. [1983] "Conservativeness and Incompleteness." *The Journal of Philosophy*. Vol. 80, No. 9 (September, 1983), 521–531.

[1997] Philosophy of Mathematics: Structure and Ontology. New York: Oxford University Press.

[2009] "We Hold These Truths to be Self-Evident: But What Do We Mean By That?" *Review of Symbolic Logic*. Vol. 2. 175–207.

[2014] Varieties of Logic. Oxford: Oxford University Press.

Shoenfield, Joseph. [1977] "The Axioms of Set Theory." In Barwise, John (ed.), *Handbook of Mathematical Logic*. Amsterdam: North-Holland, 321–344.

Sider, Theodore. [2011] Writing the Book of the World. New York: Oxford University Press.

[2021] "Equivalence". Handout for Structuralism Seminar. www.tedsider.org/teaching/structuralism_18/HO_equivalence.pdf

Simpson, Stephen. [2009] "Toward Objectivity in Mathematics." *NYU Philosophy of Mathematics Conference*. www.personal.psu.edu/t20/talks/nyu0904/nyu.pdf

Singer, Peter. [1994] "Introduction," In Singer, Peter (ed.), Ethics. Oxford: Oxford University Press, 3–21.

Sinnott-Armstrong, Walter. [2006] Moral Skepticisms. Oxford: Oxford University Press.

Stalnaker, Robert. [1996] "On What Possible Worlds Could Not Be." In Stalnaker, *Ways a World Might Be: Metaphysical and Anti-Metaphysical Essays*. Oxford: Oxford University Press.

Steiner, Mark. [1973] "Platonism and the Causal Theory of Knowledge." *The Journal of Philosophy.* Vol. 70. 57–66.

Strohminger, Margot and Juhani Yli-Vakkuri. [2017] "The Epistemology of Modality." *Analysis.* Vol. 77. 825–838.

Tait, William. [1986] "Truth and Proof: The Platonism of Mathematics." *Synthese.* Vol. 69. 341–370.

Tarski, Alfred. [1959] "What is Elementary Geometry?" In Henkin, Leon, Patrick Suppes, and Alfred Tarski (eds.), *The Axiomatic Method with Special Reference to Geometry and Physics.* Proceedings of an International Symposium held at the University of California, Berkeley, December 26, 1957–January 4, 1958 (Studies in Logic and the Foundations of Mathematics). Amsterdam: North-Holland, 16–29.

Thomasson, Amie. [2016] "What Can We Do When We Do Metaphysics?" In Giuseppina d'Oro and Soren Overgaard (eds.), *Cambridge Companion to Philosophical Methodology.* Cambridge: Cambridge University Press, 101–121.

[2020] Norms and Necessity. New York: Oxford University Press.

Tsementzis, Dimitris and Hans Halvorson. [2018] "Foundations and Philosophy." *Philosophers' Imprint.* Vol. 18. 1–15. https://quod.lib.umich.edu/cgi/p/pod/dod-idx/foundations-and-philosophy.pdf?c=phimp;idno=3521354.0018.010;format=pdf

Väänänen, Jouko. [2014] "Multiverse Set Theory and Absolutely Undecidable Propositions." In Kennedy, Juliette (ed.), *Interpreting Gödel: Critical Essays.* Cambridge: Cambridge University Press, 180–209.

Van Atten Mark, Kennedy Juliette. [2009] "Gödel's Modernism: On Set-Theoretic Incompleteness, Revisited." In Lindström, S., E. Palmgren, K. Segerberg, and V. Stoltenberg-Hansen (eds.), *Logicism, Intuitionism, and Formalism.* Synthese Library (Studies in Epistemology, Logic, Methodology, and Philosophy of Science, Volume 341). Dordrecht: Springer. https://doi.org/10.1007/978-1-4020-8926-8_15

Weaver, Nik. [2014] Forcing for Mathematicians. Singapore: World Scientific. 2010. "You Just Believe That Because . . . " *Philosophical Perspectives.* Vol. 24, No. 1. 573–615. https://doi.org/10.1111/j.1520-8583.2010.00204.x

Wetzel, Linda. [1989] "Expressions vs. Numbers." *Philosophical Topics.* Vol. 17. 173–196.

White, Morton. [1987] "A Philosophical Letter from Alfred Tarski." *The Journal of Philosophy.* Vol. 84. 28–32.

Williamson, Timothy. [2012] "Logic and Neutrality." The Stone (New York Times). May 13. https://opinionator.blogs.nytimes.com/2012/05/13/logic-and-neutrality/

[2016] "Modal Science." *Canadian Journal of Philosophy*. Vol. 46. 453–492.

[2017] "Counterpossibles in Semantics and Metaphysics." *Argumenta*. Vol. 2. 195–226. www.argumenta.org/wp-content/uploads/2017/06/2-Argumenta-22-Timothy-Williamson-Counterpossibles-in-Semantics-and-Metaphysics.pdf

Wilson, Mark. [1983] "Why Contingent Identity is Necessary." *Philosophical Studies*. Vol. 43. 301–327.

Woodin, Hugh. [2004] "Set Theory after Russell: The Journey Back to Eden." In Link, Godehard (ed.), *One Hundred Years of Russell's Paradox Mathematics, Logic, Philosophy*. New York: de Gruyter.

[2010] "Strong Axioms of Infinity and the Search for V." In Bhatia, Rajendra (ed.), *Proceedings of the International Congress of Mathematicians*, Hyderabad, August 19–27, Vol. 1. World Scientific, 504–528. http://logic.harvard.edu/EFI_Woodin_StrongAxiomsOfInfinity.pdf

Wright, Crispin and Stewart Shapiro. [2006] "All Things Indefinitely Extensible." In Rayo, Augustin and Gabriel Uzquiano (eds.), *Absolute Generality*. New York: Oxford University Press, 255–304.

Zach, Richard. [2018] "Rumfitt on Truth-Grounds, Negation, and Vagueness." *Philosophical Studies*. Vol.175, No. 8. 2079–2089.

Zeilberger, Doron. [2004] "'Real' Analysis is a Degenerate Case of Discrete Analysis." In Aulbach, Bernd, Saber N. Elaydi, and G. Ladas (eds.), *Proceedings of the Sixth International Conference on Difference Equations*. Augsburg, Germany: CRC Press, 25–35.

Acknowledgments

Thanks to Michael Harris, Juliette Kennedy, Eric Majzoub, Katja Vogt, and two anonymous referees for comments. (Shortly before submitting the final version of this manuscript to Cambridge University Press, I discovered that one of Jerrold Katz's final papers was by the same name. See Katz [2002]. The connection between the works is minimal, however. His paper only addresses the problem of Section 3 and from a characteristically distinct angle.)

Cambridge Elements ⹀

The Philosophy of Mathematics

Penelope Rush

University of Tasmania

From the time Penny Rush completed her thesis in the philosophy of mathematics (2005), she has worked continuously on themes around the realism/anti-realism divide and the nature of mathematics. Her edited collection *The Metaphysics of Logic* (Cambridge University Press, 2014), and forthcoming essay "Metaphysical Optimism" (*Philosophy Supplement*), highlight a particular interest in the idea of reality itself and curiosity and respect as important philosophical methodologies.

Stewart Shapiro

The Ohio State University

Stewart Shapiro is the O'Donnell Professor of Philosophy at The Ohio State University, a Distinguished Visiting Professor at the University of Connecticut, and a Professorial Fellow at the University of Oslo. His major works include *Foundations without Foundationalism* (1991), *Philosophy of Mathematics: Structure and Ontology* (1997), *Vagueness in Context* (2006), and *Varieties of Logic* (2014). He has taught courses in logic, philosophy of mathematics, metaphysics, epistemology, philosophy of religion, Jewish philosophy, social and political philosophy, and medical ethics.

About the Series

This Cambridge Elements series provides an extensive overview of the philosophy of mathematics in its many and varied forms. Distinguished authors will provide an up-to-date summary of the results of current research in their fields and give their own take on what they believe are the most significant debates influencing research, drawing original conclusions.

Cambridge Elements ‖

The Philosophy of Mathematics

Elements in the Series

Mathematical Structuralism
Geoffrey Hellman and Stewart Shapiro

A Concise History of Mathematics for Philosophers
John Stillwell

Mathematical Intuitionism
Carl J. Posy

Innovation and Certainty
Mark Wilson

Semantics and the Ontology of Number
Eric Snyder

Wittgenstein's Philosophy of Mathematics
Juliet Floyd

Ontology and the Foundations of Mathematics: Talking Past Each Other
Penelope Rush

Mathematics and Metaphilosophy
Justin Clarke-Doane

A full series listing is available at: www.cambridge.org/EPM

Printed in the United States
by Baker & Taylor Publisher Services